THE UNSEEN GENIUS

by
Brandon Broll

COPYRIGHT

Published by Riols Quarter Ltd.
85 Great Portland Street, London W1W 7LT, England
Company number: 12673832

A CIP catalogue record for this book is available
from the British Library

Paperback ISBN 978-1-913758-09-7

Ebook ISBN 978-1-913758-08-0

Cover design by More Visual Ltd.
Cover photo of Hermanus, South Africa,
by mlschach (iStock)
Illustrations by Barbara Jackson

DEDICATION

In *memoriam* to my father
Bill Broll
(1933-2002)
A kind, talented, gentle man
First reader of this novel

and

for J.W.

Recollections of 1986-87
when this novel was written

TERMINOLOGY

Historically a racial terminology developed in South Africa to demarcate race groups. During the apartheid period in which this novel is set, this terminology was official. As archaic and politically outdated as some of these terms are today, others are not considered offensive in South Africa (such as *Coloureds*) and remain in common use. In each case, however, I prefer to *italicize* any racialized term.

African – black inhabitant of South Africa
Afrikaans – language of Dutch origin
Afrikaner – white south African who speaks Afrikaans
Black – original native inhabitant of South Africa
Coloureds – South African of mixed descent
English-speaker – white South African, often British origin
Non-white – Black, Coloured and Indian
White – Afrikaner or English-speaking white

And as I wake, sweet music breathe
Above, about, or underneath,
Sent by some Spirit to mortals good,
Or the unseen Genius of the wood.

John Milton: *'Il Penseroso*

- CHAPTER 1 –

Sergeant Malan surveyed the coastal slope littered with rectangles of white. Below him in the shrubbery leading to the cliff-face, a few paces from where he stood, an A4-size book had unfurled its contents of white pages like a burst seedpod and was now dispersing this windborne load. It was a handwritten book torn apart by the wind, its pages strewn loosely and randomly like confetti along the rocky shore. Last night the police were alerted to this strange finding by a young woman. Sergeant Malan gazed out to his left, then turned to his right, frowning into the wind which was still dragging its early morning mist-shroud across the bay. He stood as a powerfully built man, obese with a paunch, yet proud of his rotund physique.

"Tell me Piet..." He turned to the young policeman crouching at a cardboard box beside him. This eager teenage recruit with his innocent face. "Are you keen on deskwork? Where did you learn to sort paper like that? I never realized you were so keen sorting through paper." He smiled wrily. "I could always assign you more deskwork?"

"No, Sarge..." Piet glanced up from the box, shaking his head vigorously and raising a hand in submission. "I'm happy out here doing fieldwork."

Knowing that his *protege* was a novice at this type of detective job, Sergeant Malan had assigned Piet the menial task of sorting through this papery evidence, numerically, sheet by handwritten sheet, as it was being collected by the men and had stacked it neatly in the cardboard box. Sergeant Malan now gestured to be handed a page. Piet delved into the box to find a specimen which wasn't soggy or torn or too badly smudged. At last, he extracted a dry sheet of paper that he was happy with. He handed it over to his commanding officer.

The sergeant fingered the brittle sheet and in his stubby fingers the corner tore. On the page was a text handwritten in pencil. As the wind picked up so the heavily built man turned his back to shield the page, preventing it from further tearing. The sixty A4-size pages collected so far contained a text written by the same hand, although the content on each page was obviously different. The team of police collectors on the rocky slope were still busy gathering this particular evidence, if it could be called evidence of anything more than an accident. Certainly not yet evidence of the involvement of a second party or of a struggle.

In his mind, Sergeant Malan was becoming more and more convinced that this had merely been an accident. Pages of a hand-written book, and it looked like they came from a personal journal or diary, are hardly evidence of foul play. In fact, this is what the Sergeant had assured the worried young woman last night after she telephoned in to report her discovery.

"It is not yet cause for alarm," he had explained to her with the authority of his position, recognizing the growing panic in her voice and understanding that she allegedly knew to whom this handwriting belonged. At that particular moment, the police would make a note of her telephone call in their official files and would begin an investigation in the morning.

"And Miss, you shouldn't be wandering the slopes this time of night," Sergeant Malan had warned her. "The coastal path can be very dangerous in the dark."

But apparently, disregarding this warning, in her determination and panicked worry not to wait until morning, the young woman had returned to the scene unsatisfied with this lax efficiency of the police. Because a few hours later, through her persistence and with the aid of a narrow beam of torchlight, she had stumbled around and located the bloodstained rock above the cliff-face.

Cape l'Agulhas, southernmost tip of Africa, was an hour drive from this Hermanus shore. A shore which cast its uneven line so far as the eye could see to the west, and east to the nearest buttress of eroded sandstone slanting sharply into the sea. The rock stained with blood was unexceptional in its position. It was surrounded by windswept shrubs and close to the cliff-face above the ocean waters; shrubs that were cropped and moulded into a surface as smooth as a haircut combed by the strong onshore winds. And lodged near to this rock, just above it, in a canopy of leaves below the path (these facts would later be noted with forensic precision by the police) was that ripped open handwritten book now much emptied of its content of pages.

Looking at the sheet of paper in his hand, then at the blood-stained rock, Sergeant Malan began to feel more at ease with the rock. The rock with its obvious streak of dried blood was more the type of thing police work was made of: photographing it, chiseling off samples of the bloodied stone, placing fragments into a plastic bag, sending the package off to forensics. So far it appeared only one party was involved. That there was no foul-play. In all probability it was a simple mishap. Those bloodstains most likely belonged to the author of the book. Therefore, at this stage, the case rested merely on finding the person whom, while carrying the book, had apparently slipped and injured himself and most likely fallen over into the sea.

Considering this possibility, the alarm of the young woman caller who had telephoned the police late last night was understandable. Naturally she was upset, being the girlfriend of Joseph Salem. But in this light her accusations which she raised against the police on her second and later phone call, accusing them of lax timing, inefficiency, incompetence, even police misconduct, was not only unfounded but on the face of it was quite unreasonable. In this way, assuming this attitude, Sergeant Malan reassured himself. Once the bloodstained rock had been reported the police had spun rapidly into action. After a few hours of sleep the young woman would probably see the efficiency of the police more clearly, calmly, he had thought while nodding to himself. Knowing that she was expected here on the slope soon to confirm the writing on these pages as Joseph Salem's. Then a full statement could be taken from her.

A loud exclamation from Piet distracted the Sergeant as his young recruit pointed animatedly at a black police dingy churning up a foam on the seawater, following a path to the south of where they stood. Riding on the dingy were two policemen plus a diver; they reached the first beds of kelp and offloaded the police diver who snorkeled through the dense seaweed while the lookout in the dingy scanned the perimeter of the kelp beds. That was the trouble with this papery evidence. It had become a logistical problem, being, as it was, scattered and blowing as a stream of white pages framed on the west side by stretches of beach; on the east side by outcrops of wave-beaten rock; and seawards this strange confetti was floating out to the beds of kelp, out to the fronds of seaweed swaying with the motion of the tide.

However professional Piet tried to appear as a trainee police recruit, his expression was compromised by his youthful enthusiasm. For Piet this case is what he'd always dreamed of doing. Genuine police detective work. It was his first experience of practical fieldwork assisting the team of detectives at the scene of what, in his adolescent and melodramatic conception of it, looked more and more to him like a gory and fatal incident.

Few incidents as mysterious as this happened here in Hermanus, particularly for the detective squad. The town was a small, picturesque seaside tourist resort and retirement destination on the south-western Cape coast of South Africa. The so-called *coloured* and *black* folk were generally 'well-behaved' by police standards: they had learned their standing under the apartheid system and the racially divided residential structure of Hermanus served as an effective control over crime and disorder, preventing it from spilling over into *white* areas.

Occasionally fishermen got swept off coastal rocks by a freak wave, their last places of remembrance marked by stone crosses erected along the shoreline, but that was hardly ever a detective matter to be solved. With the glut of

11

sharks in these waters, the bodies of swept away fishermen were seldom ever recovered. To Piet, compared with his first mundane six months of deskwork in the Hermanus police force, this cliff-face mystery of scattered pages was the big time.

Last night it had taken the Hermanus police team an hour to locate the bloodstained rock: its position vaguely and emotionally described by Joseph Salem's girlfriend over the telephone during her second call. Her estimation of the position of the rock was inaccurate by almost sixty metres. That didn't help temper Sergeant Malan. The men had started their search by early morning sunrise moving from one sheet of paper to the next, some pages blown a kilometer and more from the bloodstained rock. Rather than taking the girl's co-ordinates seriously since they were peppered with such angry accusations of incompetence, the police team had judged the source from which the papers had emanated from the direction of the wind. Then the red cardboard cover of the book was found with some papers still affixed into it. The rock was just a few steep steps away, close to the slide marks and uprooted vegetation, and from there a sheer drop into the sea. At this time of morning the tide was out and the water was shallower and crystal clear down to the seabed. No sign of a body. Not even a shoe or a piece of torn clothing.

Sergeant Malan turned his attention back to trying to make sense of the writing on the paper:

12

'AND THUS, IF EXPERIMENTS 34A & B ARE
TAKEN TO SUPPORT MY OVERTHROW
HYPOTHESIS IN HUMAN CONSCIOUSNESS,
WITHIN TERMS OF DARWINIAN EVOLUTION,
THE RESULTS OF CHAPTERS 1-3, 7 & 9 WILL
BE SEEN IN THEIR TRUE CONTEXT. IT IS
CLEAR THESE RESULTS CORROBORATE...'

On two occasions Sergeant Malan scratched his head. With added concentration he re-read the paragraph mouth-ing the more difficult words, slowly shaking his head from side to side. Piet's enthusiasm in suggesting that one man abseil down the cliff-face in search of the body, dissolved into silence at seeing his boss frowning at the page with such a confused expression. As Sergeant Malan repeated to himself the words 'OVERTHROW HYPOTHESIS', Piet swallowed nervously before his brimming inquisitiveness got the better of him.

"Is it subversive, Sarge?" he blurted out. "I heard a bit about Joseph Salem and the way he mixed with those *blacks*. Do you reckon he's a communist? Some men in our squad think he might be. They remember the time..."

Sergeant Malan cleared his throat loudly and eyed Piet with a stern, piercing stare. Piet stopped in mid-sentence and lowered his head, automatically clicking his heels to attention with the box of evidence at his feet. For a while both men stood facing the ocean and the sea breeze so that neither were in a position to notice the young woman approach up the path. Her arrival around a bend on the bushy winding walkway startled them.

"Miss Du Plessis..." exclaimed Sergeant Malan, drawing himself out of the difficult writing on the page. She was Joseph Salem's girlfriend, that much he did know. Of slight build and feminine, she was both attractive and mildly threatening to him in an earthy sun-drenched intellectual way with shoulder-length chestnut-coloured hair, around about Piet's age he guessed. She held a forceful presence before them, although, judging from her expression, she had obviously spent a lot of time crying. He guessed that she probably hadn't slept since those punishing telephone calls last night to the police. Her eyes looked swollen, red, and puffy. Yet still there was an aura about her of thinly disguised anger.

"Have you discovered anything else?" she questioned Sergeant Malan, her voice worried.

"Not much more since last night, Miss Du Plessis, not much more I'm afraid. But we did find this." The officer held up the red book cover. "The papers you reported blowing around the shore came from this." She noticed a dark bloodstain on the ribbing and enquired about it. In this regard Sergeant Malan pointed to the jagged rock below the path and mentioned that bloodstains had been chipped off the rock and bagged as forensic evidence. "It looks like the person to whom this book belongs, which we're assuming is Joseph Salem, slid from here and injured himself before grasping onto those branches and falling over the edge."

"Could he be badly hurt?"

"I cannot say." Sergeant Malan turned the bulk of his torso back toward the onshore breeze and stared down over the edge into the ocean.

"Sergeant, for heaven's sake, please, we've got to do something. You can't just stand here."

"Miss Du Plessis..." he said holding up his open hand with stubby fingers. "Let's be patient. You've got to give us enough time." He pointed down at the black dingy, small in its floating silhouette and distance now moving around beyond the breakers. "I have a full complement of seven men combing the shoreline on land and sea within two kilometers of this rock. Let's wait to see what they recover."

She stared beyond the dingy out to the kelp beds: those rounded floats of the seaweed at sea-level looked so much like an army of heads bobbing up and down. Was Joseph out there among them? She realized she was shivering with cold and with fear when Sergeant Malan offered Piet's services of getting her a blanket from the police truck. Although she hadn't slept all night, instinctively she felt herself recoil from this offer and turned to leave.

"No thanks. I've got to get some sleep."

"Wait a moment Miss Du Plessis, you cannot go home yet. I need you to come with me to the police station to make a statement."

"Don't tell me what to do, Sergeant!"

"Now just a minute young lady..." Sergeant Malan regarded her with intemperance. His notrils flared at this accusation which felt like almost one too many from her. "We are trying our utmost to gather evidence and locate Joseph Salem and what we need from you as soon as possible is a written, signed statement."

Some steps away down the path she paused when she realized how silly and volatile her anger seemed at a critical time like this. As she caught sight of that bloodstained rock again and the industriousness of this police activity, tears welled in her eyes. Far in the distance those kelp beds bobbed to the motion of the tide. At this moment she could hardly think straight, never mind stand up straight with fatigue, and before she could offer a reply, Sergeant Malan was leading her along the coastal path down to a yellow police pickup.

*

The first stretch of gravel road curved in a winding route hugging the contour of the land and the light police vehicle ploughed deeply into the sandy road at each bend. A subdued dust cloud rose in the wake of the van, the wheels spitting stones sideways. Already Sergeant Malan felt he had almost had more than enough of this case. Apart from the annoyance of having to start work so early on a Monday morning, this Du Plessis girl was now sitting beside him with her ungrateful, bad-mannered tongue. To distance himself from this thought, temporarily distract his mind, he leaned into each sandy bend accelerating the pickup to get as much as he could out of the winding road before its surface changed to tarmac.

Since the declaration of the nationwide State of Emergency in South Africa more than a year ago, windows on most police vehicles were fitted on the outside with a protective and reinforced casing of steel-mesh: a precaution against riots, stone-throwing, petrol bombs, unruly mobs. Together they sat as if jailed in the van, Sergeant Malan enjoying the mesh-security and twice becoming reckless in it: once at a loose bend in the road and once when passing under a low-hanging branch. He just gritted his teeth as the branch of the tree struck the windscreen mesh and set it rattling into powerful spasms. It was her first ride in a police pickup and initially she sat quietly, hands resting on her lap. But at that unexpected moment when the branch impacted the windscreen mesh, she found herself gripping onto the vinyl of the seat. Neither of them spoke throughout the journey into town.

When the Hermanus police station came into view along the straight main road, she glimpsed an old

17

fisherman whom she instantly recognized. The fisherman was being manhandled, shoved with a truncheon by an officer up the police station stairs towards the swing doors of the *non-white's* entrance. She watched as they passed through the swing doors, the fisherman disappearing down a corridor. Oblivious that the young woman beside him in the van had recognized anyone, Sergeant Malan drove the vehicle into the parking lot, pulling up outside the *white's only* entrance. By then, however, her brain had cleared momentarily of fatigue and she suddenly leapt from the van which had scarcely come to a halt, shouting:

"Stop that! Leave him alone!"

As she burst through the *non-white's* entrance doors, the officer wielding the truncheon spun round in surprise. The officer was built as solid as a farmer with hair cropped short in a crewcut and a moustache fringing a flabby upper lip. Following the cataclysmic sound of the swing doors being violently kicked open, when the officer caught sight of the cause of this disturbance: just a young *white* woman approaching angrily, he dropped his guard as she veered past to speak to the fisherman.

"Buck, are you hurt?"

The old fisherman seemed equally surprised at her intrusion but declined to answer. His weather-beaten face remained indignant in a frown of lines, as he maintained a respectful distance from the policeman holding the truncheon now by his side. Then comically, not far from

18

this young woman, the fisherman raised his chin aristocratically causing his fuzzy gray beard to ennoble his profile. She touched his sleeve. It was soaking wet.

"You're drenched," she said.

Still, he gave no reply. Unseen by the eyes of the policeman, the arrested fisherman gave her a quick conspiratorial wink, his eyes dark and secretive, then resumed his comical blank stare down the passage past the policeman. The explosion of the doors again being violently swung open preceded Sergeant Malan's entrance into the building. He called down the corridor to her.

"Miss Du Plessis, why on earth did you run ahead like that? What is your problem? Don't you understand I'm on your side, that I also want to find Joseph Salem. Anyway, this *non-white* part of the building is not meant for you."

"Reprimand this officer," she said accusingly, her nostrils flaring as she pointed a finger at the policeman standing a few steps away with a bewildered, sickly innocent look upon his face. He now held the truncheon behind his back. "Who the hell does he think he is shoving an old man in the back with a truncheon like that!" Both officers glanced at each other with knowing eye-contact. It seemed to mock her words. At this humoring of her, her anger flared. But before she could react, Sergeant Malan unexpectedly led the policeman out of ears reach a short way up the corridor.

"Hssst..." Buck signaled in a whispered tone. She was standing at an angle between the old fisherman and the two conferring policemen, strategically blocking their view of him. Curling his lip, Buck hissed once again to attract her attention and reached into a trouser pocket, drawing out a red and white chequered handkerchief. Her eyes opened wide with disbelief. It was Joseph's hanky, the one she had personally given him last year with the letter 'J' on the corner embroidered in black.

She glanced back at the two officers. They were laughing together, sharing a joke as if the truncheon incident was forgotten. With lowered voice, she nervously whispered back:

"Where did you find that?"

"Around near the seaweed beyond the waves, Missie." Buck controlled the quaver of his words by clearing his throat in a cough. "This morning I went searching for Joe. You keep it..." He thrust the handkerchief into her hand. "Missie mustn't show them." She watched helpless as he was led through a door marked *Private,* before herself being escorted out of this side of the building and into the *white's only* section.

"I'm worried about your treatment of that fisherman..." she murmured, loud enough for Sergeant Malan to overhear.

"Nothing to concern yourself about, Miss Du Plessis," he assured her, holding her arm. They passed through the charge office and into the second of several small cubicles partitioned off with misted yellow glass. "I've spoken to my man. He's very reliable. But I warn you to keep your pretty nose out of official police business. Mr Buck Williams has been detained for interrogation under section 29 of the Internal Security Act. And any obstruction by you or anyone else into this matter is a serious offence under the State of Emergency security regulations. Do you understand, Miss Du Plessis?" She remained silent as he gestured for her to sit. "We can have some privacy in here."

She sat down at a wooden table, its surface imprinted with the graffiti of scribbled telephone numbers, and watched the sergeant open a metal filing cabinet. The cubicle was cramped, barely enough space for a table and two chairs. Outside was the lightly audible conversation of policemen and the obscure shapes of people moving around as seen through misted glass. Sergeant Malan placed an official pad of paper on the table, the top two sheets separated by duplicating carbon. He unclipped a pen from his safari-suit pocket and then sat down.

"First names, please?"

"Lexa Carole."

"Date of birth?"

"12th February 1966. I'm twenty years old."

Sergeant Malan automatically jotted down her street address, knowing it already by heart as her family were locals here in Hermanus.

"Do you have your ID number on you?"

"No Sergeant. I didn't expect to come down here."

"Can you remember your ID number?"

"I'm sorry." She shook her head tiredly. "Nor do I have my book of life, OK!"

Sergeant Malan looked up from the pad of official paper. "Alright, alright. We will retrieve it later. Now I need to establish what in your opinion could have led to this accident." He removed his peaked cap and placed it on the table. She altered her position in the seat, astonished at the breadth of this question. Joseph's disappearance had happened so quickly, unfolded in such a rapid, decisive, unexpected way, she hadn't time to digest the events. Without a night's sleep to fall back on how could she be expected to think straight at this moment? They didn't even know whether Joseph was alive or not. Yet all this police officialdom and formal note taking made it seem as though he was no longer out there.

"I haven't a clue where to start," she confessed, leaning back. "What can I say?" she said, shrugging her shoulders. Looking down, she realized suddenly that Joseph's handkerchief was still clenched damp in the palm of her hand.

"Then tell me exactly what were your movements, what were your activities with Joseph Salem over this past weekend?"

"Sergeant, when I discovered those papers blowing around on the slope, I contacted you immediately. And when I located that bloodstained rock, I didn't even notice those slide marks in the bushes beside it. Or the actual book which held the pages..." Tears welled in her eyes.

"We know these facts, Miss Du Plessis. Try harder now to think back to what led up to your discovery? Delve further back. Can you think back to yesterday afternoon, to Sunday, or perhaps even to Saturday."

At that moment there was a commotion outside the cubicle. A policeman on duty shouted: "Hey, *blacks* not allowed in here! Get these two stinking *bergies* out!" There was a brief scuffle followed by the loud jumbled protestation of a drunken African voice, and the efficient sounds of a team of policemen in action. "This one's cut on the arm here: someone's stabbed him with a broken bottle." Sergeant Malan stood up, opened the cubicle door, went outside briefly and then returned mumbling: "...bloody *hotnots* drunk at this time of the morning." He closed the door. "Very sorry for the disturbance, Miss. Now you were saying..."

She wiped her eyes. "Please let me come back to do this later. I really cannot think properly."

"You know, Miss Du Plessis, I was speculating on the way driving back to the station about those bloodstains on the ribbing of that book. You have confirmed for us that the writing on those pages is indeed Joseph Salem's, and those pages could only have come from that book. But what is the explanation for blood on the ribbing? Remember we found that book nearby the path, quite a distance above the rock, which means that Joseph Salem was bleeding and injured before he struck the rock. Do you recall anything about him sustaining an injury on Sunday?"

Silently she sat with head bowed staring at her fist closed over Joseph's handkerchief. She felt exhausted. Near to her limit. And confused at this damp evidence hidden in her hand. "Sergeant, please, it's really too much for me now." She stood up, tears flowing freely down her cheeks. But she dare not use Joseph's handkerchief to dab her eyes.

"Come then..." Sergeant Malan relented, barely managing to disguise his impatience at these further obstructions from this young woman. Sooner or later, he knew, she would have to oblige the police. "I'll drive you home," he offered.

This time as they both passed down the long corridor and out of the building, an overwhelming sense of tiredness overcame Lexa Du Plessis and she noticed no further incidents in the police station.

"I think I prefer to walk home if you don't mind, thank you Sergeant," she said, remembering the nerve-racking ride here.

"Then we'll telephone you tonight. Meanwhile, Miss Du Plessis, you better get a good few hours sleep. We don't want any more foolish outbursts from you, now do we?" Without answering she watched Sergeant Malan stroll to his yellow pickup and issue instructions into the receiver of a two-way radio.

*

Outside in the police station parking lot, as the only woman standing here in the early morning light, breathing the salty seaweed air, somehow her tiredness dissipated as the urgency of the accident again overtook her. Even from this police station parking lot she could hear the sea pounding onto rocks. Listening to its distant power, inhaling the clean ocean smell, for her it again felt an indulgence to go home to sleep. More important to keep busy, useful, in touch with trying to find Joseph. So what, she thought, I hardly slept last night. Trivial with Joseph out there somewhere. How could he have just disappeared! With every retrospective image of the bloodstained rock that she conjured in her memory, she reminded herself that Joseph was an accomplished swimmer. This positivity helped. He couldn't have drowned unless he was badly injured. Thinking like this gave her a semblance of hope.

The Hermanus main road had filled with traffic. Already it was half-past eight on a Monday morning. After a few minutes of walking, Lexa stopped on the curb to properly examine Joseph's handkerchief, holding it outstretched between both hands, the corner with the finely embroidered 'J' flapping in the onshore breeze. The chequered material was almost dry. Traces of diluted blood stained it. If Buck had found this handkerchief near the kelp beds, she reckoned Joseph must have been at least that far out to sea. This type of material didn't seem likely to float a far distance in water. Folding the handkerchief with the care it deserved, she placed it back into her skirt pocket and resumed walking.

Until tears welled in her eyes. Tears associated with the responsibility she felt fingering Joseph's handkerchief in her pocket. It felt secretive. Guilty. A complex of feelings. They were tears of fear also, because she had colluded with Buck, knowing that she was hiding evidence from the police. But instinctively she knew Buck was right. She must hide it. Protect Joseph from the police. From these people trying to find him. What a responsibility. What confusion! She wondered if the handkerchief could help them find him. She dismissed the thought, holding onto his handkerchief felt right. Safe.

In such unsafe times, she thought. Thinking this way helped her better understand her volatile reaction to the police. Why she instinctively felt repelled by Sergeant Malan. A feeling she only half understood. After all, the police were genuinely trying to find Joseph. Yet these were such suspicious times in South Africa. This State of Emergency, the political uprisings all across the country. A politics of hate. Separation. Apartheid. It was obvious that the police were doing brutal things to innocent people. Suspecting communism around every corner. So much unwarranted suspicion. So much finger pointing. That it was hard to trust them. Trust them with Buck. Trust them to innocently find Joseph.

As she walked on the landward side of the main road, through this residential Hermanus neighbourhood, her fear, worry, overwhelming tiredness was met by pungent aromas from flowering shrubs leaning over garden walls onto the pavement. Behind these walls were expansive gardens, in some gardens official flags flapped in the wind. Several homes were embassy holiday cottages or government seaside mansions. Some seemed to revel in their outdated colonial architecture unhidden from the road with manicured rose gardens, pergolas, swimming pools shimmering in the sunshine. While other mansions appeared more shy in their affluence, blocked behind giant hibiscus hedges or ivy entwined tennis courts or hidden by strategically placed palm trees.

Lexa imagined the Salem family, Joseph's family, would be as worried and apprehensive as she was while waiting for latest police news. She contemplated this while passing by the lemon tang scent from an exotic shrub. Had anyone thought to check for some sort of clue to Joseph's disappearance in his bedroom? An unusual item of clothing perhaps that he may have carried off with him, or something else lying about that somehow may help to explain the accident. She walked until reaching the arrowed road-sign: *KWAAIWATER,* and here elected to bypass going home and instead veered towards the Salem household as a surge of expectancy speeded her pace.

Crossing the main road she caught sight of a police van approaching from the direction of the Salem home. As the vehicle passed, she recognized the younger teenaged policeman – Piet - who had been with Sergeant Malan near the bloodstained rock. He sat in the passenger seat with a cardboard box placed on his lap, preoccupied with reading what appeared to be a page from Joseph's book. They did not notice her. As she looked ahead from this vantage point at the entrance to *Kwaaiwater,* the avenue leading to the cove and the few visible thatch-roofed houses and the view out to sea, all appeared so calm, so unbelievably normal on the surface at a time such as this.

The Salem family home nestled above the road, facing the ocean, neatly tucked amongst dense shrubbery on the onshore slope. She approached it, knocked on the front door, and Joseph's mother opened it. As they hugged for a longer time than usual, she felt the tired slackness of his mother's grasp.

"Oh Lexa, I've prayed and prayed. What else can we do?" Slowly they rocked silently together. Then pulling apart she noticed a deep frown of concern incised on Joseph's mother's face as the woman attempted a smile, one arm held limply at her side.

"I noticed the police leaving here," Lexa said softly. "Have they discovered anything more?"

Mrs Salem gestured they had not. Arm in arm through the entrance hall they passed familiar objects in wooden wall cabinets: precious stones, marine shells, corals and a beetle collection, all gathered and catalogued by Joseph. Purposefully Mrs Salem diverted her eyes from the cabinets for the significance of their meaning to her at this time.

"I'm cooking breakfast dear? Preparing eggs, bacon, toast for us. You're most welcome."

"Coffee would be good."

Even in the kitchen amply stocked with spice racks, copper cooking pots, baking implements, a row of bottled preserves, to Lexa being in the house made Joseph's presence seem as tangible here as the smell of brewing coffee beans.

"Sergeant Malan questioned Harry and myself about Joseph's book," said Mrs Salem as she moved to empty the toaster. "He told us that Special Branch policemen are going to examine it. Doesn't that seem strange? He asked us a number of political questions: whether Joe belongs to any political organizations? Is he friendly with any black political people? And of course we said no. I can't understand what this has to do with the accident."

Lexa watched the woman attend to her kitchen duties. At these words, Lexa couldn't help thinking also of Buck Williams presently being detained in the police station. And Joseph's handkerchief in her pocket. But these things explained nothing yet and they would only serve to upset Joseph's mother by revealing them to her. Lexa remained silent. Unsure of how to reply. To her it seemed an odd kind of silence, feeling that somewhere, indirectly, a kind of politics only guessed at by her was involved. An instinctive feeling which bolstered her mistrust of Sergeant Malan and the motives of the police in their search for Joseph. But where this politics lay exactly, in relation to the accident, she wasn't sure. At this stage it had become confusing enough to know where even to start.

"Would you mind if I go up and sit in Joe's bedroom after breakfast, Mrs Salem? It will help me feel close to him." The mother turned, smiling her sympathy, immediately understanding this sharing of fears. To Lexa it felt unnecessary to explain how eager she was herself to find some sort of clue to this accident, to the cliff-face mystery, as everything seemed so vague anyway. So few facts seemed to be known. As Mrs Salem picked up and carried the tray of food out onto the porch, Lexa followed her holding a pot of brewed coffee in one hand and a jug of freshly squeezed orange juice in the other.

The porch was brightly illuminated by the early morning sun. The light shined off the sea in such a brilliant way, in bright spangles, that Lexa stood blinking, waiting for her vision to adjust. Harry Salem sat in one corner in his wheelchair with a blanket wrapped double around his legs. Surrounding him was a virtual jungle of foliage: hanging pots of asparagus and sword ferns, a well-established palm, troughs of succulent plants and a brass urn overflowing with Peace-in-the-Home. The disabled man looked tired, black rings under his eyes, yet radiated his usual confident smile. The panorama of Walker Bay lay before them through the row of porch windows. While in the next cove, hidden from this view before them, the police dingy would still be searching among the kelp beds and wave-beaten outcrops of rock for Joseph. The scene of the accident was just that close.

"Morning Mr Salem."

He nodded an acknowledgement to Lexa's greeting, inclining his body forward in the wheelchair to straighten up and receive his breakfast tray.

"We had the police around here a minute ago," he said to her. "They asked us about those papers blowing on the slope around *Voelklip.* Joe's handwriting was recognizable, of course, but neither Hermine nor myself have any real idea what it means. Apart from knowing about his hobby with animals. Did he mention any details of this book to you, Lexa?"

Mrs Salem handed Lexa an earthenware mug filled with coffee as she perched on the windowsill, listening. For some inconspicuous moments Lexa brooded on why Joseph had sworn her to silence on this issue. She was beginning to feel the impossibility of remaining reticent and secretive about this at such a time. Notwithstanding that he had simply asked her not to tell anyone, in these circumstances it was becoming too great a burden to withhold all information on Joseph's account.

"He was working on a long-term project," Lexa began hesitantly, with a degree of caution, but then fell into a silence struggling with herself, with the ambivalence of knowing that something must be said but how much? It was true that more than anyone, so she believed, Joseph had confided in her over these matters. To the extent of otherwise being hyper-protective of his privacy. He hardly confided in anyone. Over the years he had begun to trust sharing with her these science writings some of which she only partly understood, others of which she was happy just to assist him to complete. Certain theories he was developing involved sensitive subjects which bordered on human ethics and the genetics of human behaviour. With his heightened sense of scientific responsibility Joseph mostly confined these issues to intensely private study.

"Joe hinted to me that he'd discovered his life's work," Lexa said, somewhat embarrassed. But just as soon as she had blurted this out, she stopped because of how trite it sounded, how boastful, as if she was relinquishing his privacy with cheap words.

Mr Salem eyed her. "Life's work? You mean he was writing a book about it?"

"Yes, exactly!" she confirmed now with clear conviction in her voice. "He believed he was unravelling some sort of code to understanding human behaviour in a new way. Using a new scientific method in the way we see ourselves. Often, he would fall into deep and silent contemplation about it. Once when I asked, he said he was weighing up one theory against another. I helped him to design experiments. He carried a little notebook in his shirt pocket and would reach for it at any available moment, immersing himself in concentrated bouts of writing. This happened just about anywhere: on buses, in the cinema, secretly at our kitchen table, even when he was doing his seashore surveys. He admitted once that he kept the little notebook beside his bed at night and would wake up with an important idea and need to jot it down. But it was just a small pocket notebook. Not the large book on the slope found by the police."

"Ahh..." Mr Salem nodded. "You mean the book the police discovered is connected with those notes he used to scribble down?"

"Yes, I think so," said Lexa.

"Didn't we used to ask Joseph about his note writing, Harry?" asked Mrs Salem. "Yes!" she smiled with amusement. "What are you writing down son, I used to ask him. Are you listening to us? I would inquire when he seemed so far away with his thoughts. But you know Joe with his good humour: laughing it off, always keeping very private about the nature of his hobbies."

Mr Salem shifted his body weight on muscular arms. The wheelchair squeaked with the motion, as his vision scanned across Walker Bay. There were still wisps of mist running low on the water in streamers along the horizon. The thick muscles in his jaw bunched.

"Well that's over now," he said.

"What do you mean, Harry?" Mrs Salem protested.

"No!" he reassured her quickly. "No, I'm sure he's safe Hermine. Let's not become emotional about that or jump to any conclusions that Joseph is hurt. I'm talking about those political notes, those pages blown all over the place. How many times did I warn him to avoid Buck Williams and the others. Mix with his own kind: not get involved with those *Coloureds*. Keep away from *black* politics. Not that I've got anything against *Blacks*, of course not. It's just that in these dangerous times with the countrywide State of Emergency you don't go playing around with fire."

"Mr Salem, may I please interrupt."

His face was frowning at the expanse of the sea.

"As far as I know Joe was not directly involved in anything political. His interests were always scientific, they were in biology: that's something we all know. I really cannot believe that those pages the police have collected are anything other than scientific writing."

"Why are the police so worried then?"

"I can't exactly say I trust them with this," Lexa mumbled under her breath.

"Pardon?" Mr Salem's gaze remained fixed beyond the windows. He didn't look around.

"Harry, I'm sure they've made a mistake." Mrs Salem smiled while struggling to retain her composure. She watched her husband glaring sternly out to sea. At times like these, when her disabled husband seemed upset or angry or disappointed, she believed he needed her most. She began to butter him a slice of wholemeal toast.

Lexa watched them. In particular, she tried to gauge the expression on Mr Salem's face, remembering the greeting he had affected when she first stepped onto the porch. Somehow his smile of greeting seemed unlike the smile of a man whose son's life was in the balance; it seemed under the surface almost to have a subtle tinge of victory stamped onto it. Could that be so, she wondered?

Or were her suspicions: of the police, now of Mr Salem, starting to get out of hand. No, she thought, I am mistaken. But how could a father consider in the cold light of reasoning his son being involved in politics at such a critical time as this, and then make a wild accusation about the ending of Joseph's life's work. Lexa felt herself beginning to wilt under the strain and tension of all these thoughts. Leaving half her mug of coffee undrunk, she excused herself from the porch, and mounted the flight of stairs up to Joseph's loft bedroom.

*

She remembered Joseph once curl his lips in wry amusement at the statement:

"Do you know what Einstein's third law is? (knowing of course there was no such law). Neatness is relative!" He'd offered this joke describing the unusual state of his loft bedroom whilst shrugging off any criticism: "If I can locate everything, why make such a big deal about the enormous clutter that is my living space?"

Now, inwardly, she smiled at this memory of Einstein's third law and the laughter they'd shared as she entered the chaos of his laboratory bedroom.

Feeling the unmistakable presence of Joseph in the room, Lexa's eyes filled again with tears. She paused, fighting back waves of emotion welling up, her sense of sadness, emptiness, the fear for Joseph's safety. Then she approached a large pin-board on an opposite wall beneath a sloping roof window, hoping to find photographs of him, but there were none. Not even a single picture. At this dreadful moment when she needed to console herself with a portrait of him, it seemed ironic to confront photographs of herself on this pin-board. Two photos revealed her holding up various plant specimens to the camera; another showed just her own recognizable fingers clasping a small marine crustacean held next to a centimetre scale. And yet in another she was standing further away from the camera at a transect pole helping to measure out an experimental plot of indigenous plants. Then she recognized Buck Williams in a photo cradling an injured Jackass penguin in a rowing boat. Invariably, however, none of the people in these pinned up images, even Lexa herself, were the subject of the camera's roving eye. All of the people were portrayed more as willing props who were assisting Joseph in his studies of natural history.

Knowing Joseph as well as she did, each photo was understandable in its importance to him. Uniquely these were Joseph's priorities. As she walked around the room she ducked under sampling equipment and instruments he used for scientific monitoring as well as scuba diving gear suspended from the ceiling like so many mobiles. Scanning the room, she stopped, and the silence of his presence throbbed with her heartbeat. Lexa sat on his bed wondering where a clue may be found amongst all of this unusual and diverse paraphernalia. Then she remembered the hand-kerchief inscribed with the letter 'J' still in her pocket, and drew out of her other pocket the folded title page to Joseph's book:

A SCIENTIFIC MODEL ON THE EVOLUTION
OF HUMAN CONSCIOUSNESS
by J.W. Salem

By sheer coincidence this title page to his book happened to be the first piece of paper she had extracted from under a coastal shrub among dozens littering the slope last night. Recognizing his handwriting which was confirmed by Joseph's name in the title, she had quickly raised the alarm by telephoning the police. The coincidence of finding this title page seemed like a miracle, a secret gift to her alone, so had she thought at the time, and determined that the policemen were not going to claim this title page from her, or even know about it. Much like she felt about the handkerchief now. Anyway, she reckoned, they had collected more than enough pages from the book themselves to choose from.

Although just over a day had lapsed since Joseph's disappearance, the time had drawn on interminably since the telephone call and the waiting seemed so much longer. She closed her tired eyes to this thought. She took a deep breath. Sighed to herself. Surely a day adrift in the sea didn't mean it was too late to find him alive? Apparently, he took a walk alone after his 21st birthday party on Saturday night: no one thought anything of it. Frequently he went out at odd hours observing some natural phenomenon or monitoring the changes in marine life at low tide. But what made him go out carrying that bulky book? He never returned for the rest of Saturday night, nor the next day, and seldom slept away from home without informing someone. That dreaded six o'clock moment on Sunday evening had imprinted itself on her memory: when she visited the Salem household quite innocently, unsuspecting of not finding him in his bedroom, pieced together this set of events, retraced Joseph's footsteps and found the papers scattered along the shore.

Slowly scanning the sweep of his bedroom, a second visual inspection of his possessions, did not help her establish what kind of clue she should be looking for. Nothing in particular stood out of place. Unusual objects affirmed their presence everywhere in this room, out of the ordinary scientific collections, each of which reflected one or other aspect of Joseph's personality. When she raised her tired gaze along the sloping angles that made up the loft ceiling, her reddened eyes strained at the bright morning light streaming in from roof windows.

Lexa shifted herself into a more comfortable position by leaning back and settling herself amongst pillows on the bed. In doing this her eyes were distracted, fixing on a poster which occupied a vantage point on the angled ceiling above the bed and which could be viewed lying down. The poster was an enlarged photograph showing an encounter between a leopard and a baboon on a desert salt pan. It was a real life-or-death encounter which Lexa finally closed her eyes to, and to the muffled sound of waves and the onshore wind outside, she fell asleep.

– CHAPTER 2 –

The African cat in the poster crouched, poised, ears raised gundog-style. Its graceful feline features were masked up to its eyes by the brown grass. The leopard knew of the dry crispness of the grass, that each noisy step was a risk to its concealment. It knew that the death strike, when it came, must be exact and firm on the throat; choking by strangulation and the severing of that vital jugular were one and the same, so long as the action was decisive and quick. That way the leopard was best insured against injury by the flailing limbs of its prey. These measures glimmered in those green searching eyes. A subtle frown of experience was taking hold of the cat's patience: its chance was approaching; it would be soon...

Fifty paces away the troop of baboons neared a clearing onto the salt pan. These primates had been feeding off ripe figs which had dropped from heavily laden trees onto the dusty ground. The bellies of two frolicsome juveniles were distended in a sort of primate cuteness; three of the adults were starting to feel the effects of the natural fermentation process of alcohol in the figs they were eating. The relentless mid-morning sun blazed down on them. Deep cracks on the pan were widening as crystals of salt glistened in the haze of rising heat.

Proudly a sub-adult baboon ventured towards the perimeter of the troop. After having groomed one of the dominant males, it carried itself with a stature of confidence. As the acknowledged expert in the troop at catching the desert locust, this adolescent baboon often exchanged half-eaten insect morsels that it had caught for favours, especially for the status of travelling close to the troop leader. When locusts were plentiful this young male was favoured to share its prodigious catch with up to one third of the baboon troop. But today, the dizziness brought on by the alcohol-laden figs was inducing a recklessness to its confidence.

A brown speckled locust waggled its antennae and adjusted its perch on a grass tussock bordering the salt pan. One of the juvenile members of the troop had disturbed the large insect whilst romping in the shade of an *acacia* tree, and the 'click-click' buzz of the insect's wing when taking off in flight aroused, instinctively, the locust-catcher's ear and eye. Four less-than-steady bounds placed the young catcher baboon within arms reach of the grass tussock, but by this time the insect rose into the hot air over the pan. And the prodigal baboon realized, just a moment too late, that its intoxicated momentum had deposited it also out onto the salt pan. Out into the open. Completely alone.

The leopard surged forward. The crouched rump swaying of the cat and its gentle hindleg kneading on the sand was, in one graceful movement, transformed into a deadly arc of motion. The cat burst smoothly, stream-lined like a projectile, through the grass. At close range its spotted coat was a blur of contrast, but at forty paces out on the salt pan the baboon first saw the cat in a frozen charge of spots.

Time enough for its primate reflexes and its growing terror to lead to a reaction. There was no chance of running back towards the barking troop as the troop members retreated, barking their aggression at the leopard, howling their group warnings in their own drunken stupour. The only defence for the young baboon in this space of milli-seconds was to face the feline predator head-on. Courageously. Alone. As hackles raised along its spine, the prodigal baboon stood fast, its toes digging into the cracked salt on the ground, as it met the cat with a grimace of pointed canines and foam-laden screeches...

Through the dust cloud raised by this commotion of near death, Joseph Salem blinked from this daydream and lowered his eyes from the leopard-baboon poster on his bedroom ceiling. He lay languid on his bed. It was Friday morning. Three days before the police would begin their search for him above the cliff-face. He was distracted from this poster scene, from his creative reverie of what may have led to this salt pan encounter and whether the baboon survived, by his mother announcing from downstairs that the family breakfast was almost ready.

*

Sunrise this morning, to the delight of Joseph, had protruded its head as a laser fork of rays beneath the heavy gray-blue of passing storm clouds. The band of shrubbery in which the *Kwaaiwater* homes nestled resembled a dense indigenous patchwork of green hazed this morning with sea spray. Next to *Rus en Vrede*, their seaside cottage, were a few Cape-Dutch style homes on the shoreward perimeter of *Kwaaiwater.*

Many inhabitants of *Kwaaiwater* were retirees from Cape Town, the largest city to the west in the province of the Western Cape. Apart from accommodating retirees the town of Hermanus also played host to the whims of Capetonian holidaymakers who, after an hour and a half drive by car from the metropolis found themselves within the clutches of a rugged, unspoilt coastline. It was the peace and quiet and the rugged natural beauty of the place which had attracted the Salem family to settle along this shore.

A barking dog could be heard outside Joseph's bedroom, its excited yelps discerned through an open sea window as the dog passed by the house on its way down to the coastal path leading to the beach. The dog sounded like Buck Williams' Labrador. Joseph half expected to hear the old fisherman calling after the animal but instead heard a faint whistling of a recognizable tune: *'I was born under a wandering star.'*

Joseph smiled to himself. Years back in Hermanus' Guest House Hotel he recalled smuggling Buck into the darkened dining room where an audience of *white's only* members were watching a rerun of the 1970's Hollywood film 'Paint your Wagon'. The film starred Lee Marvin and Clint Eastwood as two maverick gold miners, a subject that went down a treat in the mining psyche of many South Africans. Ever since Buck's first viewing of this musical, it was as if gold miners and fishermen shared a secret pact. And Buck had latched onto that catchy tune: 'I was b*orn under a wandering star.'*

Melodically he whistled it as if the tune struck a deep personal chord with him, repeating this simple phrase over and over *ad nauseum,* the rest of the song only half known and in places corrupted by hoots of laughter (plus an odd word of sympathy for the two maverick gold miners), before again piping in with the punch-line: *'I was born under a wandering star.'*

Joseph rose from his bed. He approached the open loft window facing the sea, but coastal shrubbery obscured any view of Buck or his dog. He thought of the benefits of a word of wisdom from Buck. A serious issue was weighing on his mind which needed to be resolved this weekend. In the distance the curling waves thumping and splitting onto jagged rocks, rows upon rows of heavy dumpers, had a haunting yet passifying sound as he gazed out at them. Today the picturesque cove was living up to its name: those waters sounded angry, indeed they did sound '*kwaai*', although perhaps amplified by thoughts that were weighing on Joseph's mind.

"Breakfast is ready!" his mother repeated her call.

Downstairs Mr Salem sat at the far end of the breakfast table in his wheelchair. As Joseph entered the dining room, his father appeared disembodied, with only his hands visible holding up *THE ARGUS* newspaper, Friday 5th December 1986, with bold headlines:

VOTES FOR ALL IN S.A. - Call by Shultz

Washington: United States Secretary of State Mr. George Shultz has called for universal franchise in South Africa in his first detailed attempt to define what should replace apartheid. Mr. Shultz promised firm American support for South Africa's whites if they started a dialogue with other groups to find a system that would include constitutionally guaranteed rights for all.

U.S. 'MUST CLARIFY STAND ON TERRORISM'

Pretoria: The US Secretary of State, Mr. George

> Shultz, and his Government would soon have to decide where they stood in their struggle against communism and terrorism, the South African Minister of Foreign Affairs, Mr. Pik Botha, said today. Mr. Botha was reacting to a speech made by Mr. Shultz in Washington yesterday in which he urged South Africans to begin talks on a "multiracial democracy", warning that time was running out for peaceful change.

Joseph took scant notice of this political reportage as he circled the table.

"Morning son," Mr Salem folded the broadsheet, shaking his head in wry disbelief, a bemused expression on his face. Joseph pulled out a seat and sat down. "Bloody Americans," Mr Salem continued, "thinking they can prescribe how we should run our country. I agree with Pik Botha: communism and terrorism are the real problems here in South Africa."

"Excuse me?" asked Joseph.

"Newspaper headlines son. You know... the so-called 'international community' interfering in our political affairs again."

"Oh, that."

"Apartheid, apartheid, they forever bemoan. It's just not that simple. Black South Africans in this country have a lot of growing up still to do. You okay son? You look a bit down."

"It's nothing really."

Just then Mr Salem glimpsed out the side of his vision Buck Williams coming into view through some shrubbery on the coastal path. Although the old fisherman was a distance away, even now the onshore breeze carried emanations of that whistled tune: "I *was born under a wandering star.'* But the tune was now whimsically being played on a harmonica.

"Is that Buck?" asked Mr Salem.

Joseph turned around to face the sea. He nodded.

"If you per chance bump into him this weekend, Joe, tell Buck he still owes me for the fishing rod." Mr Salem confirmed, placing a heavy hand on the wheel of his wheelchair. "He knows I can't come asking for money."

Together they watched the fisherman finally reach the beach.

"Let me pay you for what Buck owes, Dad. He's been hard up for money for months now. You know that."

"No, it's not your affair. It's a point of principle. Anyway Buck promised me he'd pay me back and I took him at his word. That's important."

"Alright. I'll remind him."

Observing the fisherman in his ragged clothes, Joseph felt a twinge of guilt. No longer playing the harmonica, Buck was throwing lengths of dried brown seaweed for his dog to retrieve.

Mr Salem's eyes warmed to Joseph and he smiled at his son. One hand touched the tablecloth near to Joseph's place setting.

"Joe, I'm getting a bit worried about your movements, son."

"Why?" asked Joseph.

"That short cut you take to Fernkloof Nature Reserve. I'm concerned it's not safe any longer. In these unusual times of the State of Emergency, you shouldn't be passing through Hermanus *Coloured* township."

"But it's the only short cut, Dad."

"I understand that, Joe." Mr Salem's nodded, then his face became serious as his index finger rested on the text of the newspaper on which violent incidents of political unrest were listed. "Problem is it's here in black and white. These times are turning really dangerous now. So Wouter Thyssen kindly offered to keep an eye out for me to make sure you don't pass through his property on the way."

"But you've never done this before, Dad."

"I've never had to do it son. Don't blame me. It's not my fault. I worry about you. Let's hope the times change back and the situation becomes normal again soon."

Joseph's expression remained puzzled.

"How did it come to this? I don't understand. You've just restricted my movements and told Mr Thyssen to watch out for me. Just a moment ago you were talking about the Americans."

49

"I was reading about communism and terrorism in our country. As stated by Pik Botha, our Foreign Secretary. And the dangers involved."

"Do you think that Hermanus *Coloured* Quarter is full of communists and terrorists?"

The man shrugged. "We can't afford to take a chance."

"No way. I know those Hermanus people. And they know me. Dad, when last did you visit Cape Town? Driving back here yesterday in the bus from university we passed Crossroads, you know that shanty town on the city outskirts, much larger than the Hermanus *Coloured* Quarter. And no way, neither can all those poor Africans living there be communists or terrorists. But I tell you: Crossroads has become a war zone. It's so desperately sad. It's half burnt to the ground. The army has cordoned it off with razor wire; armed vehicles patrol the perimeter. I've spoken to university colleagues who conducted research there and they say the army isn't interested in stopping black violence inside the razor wire."

"If blacks want to kill themselves, that's their business. It proves my point. I won't stop them."

"But Dad, it's not simply blacks killing blacks. A professor of politics gave a talk at the university providing evidence showing that the apartheid government pays teams of black assassins to kill people who oppose apartheid. In fact, I even read a report that the army used flame-throwers to set alight the shanties in Crossroads, fomenting more violence..."

"What utter tosh!" Mr Salem exclaimed, laughing at this apparent twisting of the facts. "I don't know what they teach you at that university. But that's tosh. Rubbish. War is war, bombs are bombs, and I say don't spare those terrorist bastards."

"Women and children too, Dad?"

"Hells bells, grow up Joseph, you're such a baby! Now pass me my pipe tobacco on the bookshelf over there and I won't hear any more of this." The broad-shouldered bearded man watched Joseph's departure with a degree of agitation, disappointment, disrespectful of the views which his son held. Views which Joseph seemed to be making more forthright these days. He leaned back and began to finger the spokes of his wheelchair. It was strange how the spokes of these wheels brought him comfort. A sense of peace. Calm. The family often joked about this wheelchair habit that made this burly immobile man rock back and forth. Before his accident Mr Salem was a physically active farmer and this rocking of his wheelchair had replaced the energetic remnants of those operational days.

By the time Joseph returned and duly handed over his father's pouch of tobacco, Jessica, his younger sister, had seated herself and warm scones were placed on each side-plate. Mrs Salem followed the maid as she brought in the boiled eggs nestling in egg cups, then a tray of coffee, before the family commenced eating.

As they ate breakfast, Mr Salem listened intently to Jessica discuss the merits of a subject she was studying at school and a teacher involved in this subject, before clearing his throat and turning to Joseph.

"Well, it's your weekend, son." Nodding thoughtfully, Mr Salem smiled across the table while spreading butter onto a scone. "Your birthday tomorrow. You'll be twenty one. I still can't quite believe it. How fast time has flown, and how we've waited so long for it. The time finally to commit to your birthright, to make your final decision."

Joseph raised an eyebrow.

"Harry please!" interjected Mrs Salem in a reasoning tone as she overheard. "Let's leave this issue alone over breakfast as we've discussed it enough times."

"Alright Hermine," he smiled. "That's fine son." It was impossible not to see how much thought the man had invested in what he termed 'Joseph's birthright'. Mr Salem raised his hand as a 'toast' holding a scone topped with marmalade. "This is just a reminder Joe – my only son - soon to be twenty one," he announced proudly with fatherly sentiment and enthusiasm. "The official papers wait for you over there in the oak bureau. Next Monday afternoon Mr. Rademan will visit the farm. When you meet our attorney at two o'clock with the official papers, he will sign you up, and Uncle Jack will witness. Then my brother won't doubt that we're back in business at *Salemkop*."

Legally, as a 'birthright' and now by age, on Monday Joseph could assume his place as joint manager of *Salemkop*: the Hanepoort wine farm the Salem family had owned for four generations. Overlooking the majestic Simonsberg mountain range, this established wine farm of modest size was popular with tourists investigating the Cape wine route. This weekend, his twenty first birthday and this inheritance were inextricably linked. Although Mr Salem knew Joseph felt uncertain taking up this inheritance, he reasoned it was a necessary uncertainty to be overcome by his son when relinquishing childish expectations of his youth. It was all about stepping into adulthood. Taking adult responsibility. Taking up his Salem birthright. Mr Salem's penetrating and determined blue eyes were proudly fixed on his son from across the breakfast table as Joseph avoided meeting this gaze.

Jessica and Joseph cleared the table after breakfast while their mother wheeled her husband into the enclosed porch. Then as a brother-sister team, Joseph soaped and scrubbed the used plates and dishes as Jessica dried and stacked them onto a wooden rack, talking in her incessant schoolgirl way, at times resting a hand on her brother's shoulder, her touch a gesture of solidarity, sisterly love, glancing at Joseph with a caring which seemed older than her years. Although younger her outlook was wise and mature, but being female he knew she was unfairly excluded from any farm inheritance. Jessica's close sisterly bond with Joseph helped her to understand in her own way the kind of personal compromises Joseph would struggle to make were he to take up the inheritance.

*

Releasing the brake of the wheelchair after wheeling her husband from the breakfast table onto the porch, Mrs Salem covered his crooked legs with a tweed blanket. Turning her back on him, she adamantly fixed her gaze on the windowsill of the porch as though the window and bright ocean world outside didn't exist. Folding her arms, she stood looking downwards for some moments.

"How could you raise the subject again, Harry?"

"It's not a matter of whether I can raise the subject, dear. You know my feelings. Joseph must be directed as to what is best for him and his future. Even if I have to mention it every breakfast he spends with us in Hermanus. Anyway, I know I'm vindicated..."

Facing south and swathed in morning sun, Mr Salem reached for a thick leatherbound book on the coffee table beside him. He opened it at an embossed bookmark of parchment and burgundy-coloured silk. "Solomon 29-16, I think..." He quickly flipped through four pages beyond the bookmark, then ran an index finger down columns of script. "No. I beg your pardon. Proverb 29-17. Hermine dear, consider this:

> *"Discipline your son, and he will give you rest;*
> *he will give delight to your heart."*

Slipping the bookmark into its newfound place he snapped the bible shut, running his palm over the stippled leather cover as if he were again a farmer stroking the hide of a live animal.

"A more pertinent proverb you couldn't find".

Mrs Salem shifted her weight onto her other foot. She remained silent, arms still folded across her breasts. An old feeling welled up inside for an instant. That uncomfortable, inexplicable surge of something she couldn't really describe or easily put a name to. Frustration? Anger? No, I mustn't think like that, she scolded herself inwardly. It felt like a surge of pent-up energy but surely, she reasoned, surely energy couldn't be dampened that quickly. And anyway energy for what? Her energy was spent on Harry: he needed it more than her, consumed it like a battery, and she felt honoured to provide it. In fact, desired it. Desired his acknowledgment of her. She sensed that few couples nowadays seemed as devoted to one another as Harry and her: body and soul, like a clockwork mechanism. She moved faster than him like the minute hand, and he in his directive way pointed out the hours and paced her, guided her with a curious uplifting rationality which she couldn't often fault.

Years ago this quality had made him one of the most respected wine farmers in the Paarl District. Locally famous. Agriculturally productive. Who could argue with that? Who could argue with quotations from the bible, as her husband interpreted them? They seemed just so apt. Ultimately all she could correct was his manner of speaking to her or the children. When he became abrupt, worked up, frustrated, often due to his disabled state, she was able to offer this guidance. Over the years she concluded that it was not really what he said to her or to Jessica or Joseph, but how it was said that mattered. In this way she tried to instill softness in him. Thinking about it in this way, when she rationalized it, she found that that twinge of uncomfortable 'energy' rather than anger, subsided.

"Harry..." Mrs Salem's voice emerged somewhat milder, more withdrawn than expected, as she turned from the porch windows to face her husband. "I suppose in the longer-term Joseph will come to appreciate your advice. He does love you. I know he does. But please dear, try to be gentle on him when you speak your mind."

*

Jessica possessed a copious mane of dark hair and features somewhat Middle Eastern. An intelligent seventeen year old completing her final year of school examinations, her chosen career was in music, all the family knew this and she desired it more than anything. As musical and sensitive as she was, her face this morning revealed her concern about the events at the breakfast table between Joseph and her father. Whenever she felt disturbed by family matters, Jessica had a habit of raking her auburn hair with her fingers and flicking it over her shoulders. Joseph glimpsed the movement, recognized it across the kitchen as he placed a few bottles of condiments back into a kitchen cupboard.

Her love for her brother, her respect for his talents and his dedication which she understood in herself in her music, engendered a closeness between siblings that was often commented upon by outsiders. In this closeness Jessica flattered herself that she understood Joseph well. Knew what made him tick. But his years at university away from home had added a dimension to his moods that at times were all too serious to fathom. Such hidden, such un-manageably hidden and secret thoughts now seemed to weigh in his eyes.

Unsure how to broach Joseph's mood as they completed the drying and putting away of breakfast cutlery, it was enough, Jessica thought, for her to chat lightly, uncontroversially, about what had transpired in her life over this past fortnight and to lean a comforting elbow every now and then on Joseph's shoulder. Not realizing in the process how she was frowning whilst gently raking at her hair. Nor that Joseph had sensed her reaction to comfort him, and in turn was himself searching for something which would distract and lighten her frown.

"Watch Jess..." he whispered, motioning with a finger to his lips for her to keep silent and still.

A butterfly had flown into the kitchen from the garden over the open upper stable-door and flitted around for a while, before perching on the handle of a broom. Joseph sank to his hands and knees.

And there before Jessica's eyes which widened in amusement, Joseph's crouched movements transformed into the motion of a chameleon. Mimicking in a hilarious yet scientifically accurate way the gradual locomotion of the chameleon, a backward-and-forward swaying on hands and knees, a motion that required great reptile deliberation and control. Holding back her urge to laugh, she remained transfixed on the strangeness of a brother who had taught himself this through hours of observation, the patience of the chameleon in human form. Motions which resembled the flutter of a leaf lightly blown. With the degree of care he invested into this very demanding effort, it took Joseph a couple of minutes to cross the floor: in a steady swaying progress...

Reaching his goal, Joseph's hand began to stretch out towards the colourful insect as Jessica's eyes alternated between the butterfly versus her brother. And then to her astonishment the butterfly gently mounted his finger, pumping its proboscis and swaying two antennae as though Joseph's finger were a branch or a flower of the most natural design.

*

"Tell me your plans for this weekend son?" Mr Salem enquired with added effort at politeness, as Joseph joined him on the sunny porch.

His father's friendlier tone of voice eased their meeting. The man sitting amongst potted plants smiled a greeting which was passive now. That earlier political disagreement and discussion around limiting Joseph's movements in Hermanus seemed to have dissipated. Joseph also hoped the farm inheritance issue was, for the moment at least, banished as well. This obvious mood transformation in his father was probably the work of his mother, Joseph guessed. Sometimes her motherly reprimands had this cooling effect for which the rest of the family were grateful. Joseph leaned against the glassed-in balcony, his back facing away from the sea.

"Well, no more studying for exams," he said.

"That must be a relief. I can't wait to see the photographs of your graduation, Joe. I only wish I could have joined you had there been wheelchair access. Seems hard to believe my son was awarded all those class medals and now you're a graduate in science with an Honours degree in biology."

Joseph modestly kicked at the floor.

"Thanks Dad."

"When are you moving out of university residence?"

"I pack my things tomorrow."

Mr Salem nodded a congratulatory nod, pausing for thought, while enjoying the pleasurable December summer sun already radiating its heat at this early time of morning. The pale blue sky seemed endless in its expanse with a few lone, long streamers of cloud moving almost imperceptibly across the bay.

"Tell me about your fortnight field trip after the exams. Your field trip to Namibia."

Joseph shifted his weight on the windowsill.

"Exams are exams! There's little to say about university exams, Dad, except I couldn't have done too badly since I graduated with distinction."

"That's my boy. And what about Namibia. How was it?" Mr Salem's expression was genuinely enthusiastic, in a way he couldn't or didn't attempt to hide the intense interest he held in subjects of Africa.

"One moment," said Joseph, raising his finger. "Jessica also asked to hear about the Namibia field-trip." Joseph departed and returned shortly with Jessica who seated herself next to her father on a chair amongst the potted sword ferns.

"I don't exactly know where to start?" Joseph paused for thought.

"Start from the beginning," said Jessica.

"Ok. Professor Harris and myself, just the two of us, flew north to Windhoek where a driver picked us up in a land rover *en route* to the Gobabeb Desert Research Station. Do you remember that I went there previously to complete my Honours project on the desert chameleon?"

Mr Salem nodded, remembering clearly. How could he forget? Joseph's absence had meant that, for the first time ever, he was unable to attend annual review duties at the farm at *Salemkop*. The crucial training for a farm manager this time involved a viticulture expert from upcountry assessing their vineyards, offering advice on everything from cellar maintenance to supervision of the *coloured* labourers.

"The Namib desert", Joseph continued animatedly, "is actually made up of two deserts separated by the Kuiseb river: which is a parched, dry riverbed. North of the Kuiseb the desert is gravel plains: a vast flat expanse of coarse stony ground almost featureless in the shimmering heat haze except for random sun-bleached tree-stumps which throw long jagged shadows on the ground. Passing through these plains by land rover, our driver, Professor Harris and myself, tracked a herd of gemsbok antelope slowly ambling to a waterhole. We tasted the saline water at the waterhole but found it undrinkable. Bitingly salty. In fact, encrusted with salts. Yet the antelope had trudged mile-after-mile just to lap at the precious wetness.

South of the Kuiseb river is the sandy desert: a great shifting sea of sand, mostly rust-tinted, but variously coloured from yellow to dark red to ochre. Massive dunes - some of the tallest in the world - continually moving and transforming as the wind moulds their shapes. The channeled winds throw plumes like red smoke off the top of dune ridges. Then sand cascades with a stinging speed down the sides. The dune patterns are never the same, undulating to different shapes daily."

Mr Salem bit at the grey hairs of his beard. His eyes showed an almost glazed look of fascination carried away imagining a new dimension to this continent. It was the Africa he loved.

Meanwhile Jessica sat slightly bemused, smiling at the way Joseph waxed lyrical whenever he described his field trips.

"The Gobabeb Desert Research Station is situated on a bank of this dry Kuiseb riverbed. Strategically they built it at this junction of both these deserts so that after a five-minute walk from the gravel plains, you're suddenly in sandy desert climbing up the dunes. For a full week I visited my chameleon study site in the south west, trekking each day with electronic equipment: monitors, heat sensors. And from sun-up to dusk (protected by a makeshift canopy at the hottest time of the day) I ran my experiments amongst clumps of *Stipagrostis* grass to measure the ambient temperature – of ground and air -

and the amazing body temperature of the Namaqua chameleon."

"Where was Professor Harris?" asked Jessica.

"He visited my study site a few times, offering advice, standing in to collect data for short periods, but he was also busy supervising other scientific projects at the research station."

"Anyone for tea?" Mrs Salem stepped onto the porch carrying a tea tray with biscuits on a plate. Shortly she departed mentioning that the maid needed supervision with some fruit preserves.

"The aim of my Bachelor of Science Honours project was to establish how the Namaqua chameleon stays alive in the baking heat of the desert."

Jessica leaned forward enthusiastically: "What did you find out?"

"It's quite amazing. Are you listening, Dad? We discovered that the Namaqua chameleon controls the temperature of its body by altering its skin colour. The many experiments I conducted proved, at least in the Namaqua chameleon, that any matching of its skin colour to its background as though the animal were camouflaging itself, is in the desert more a coincidence of the animal trying to control its body temperature so as not to overheat."

"How so?" Jessica frowned.

"You see, being a cold-blooded reptile the chameleon needs to warm its muscles each day to get ready for action much like an insect. But in the desert this chameleon doesn't merely warm up, its blood can just as easily boil. Reptiles can't produce their own body heat nor cool themselves down internally like birds or mammals, so they depend on the environment. Take a crocodile for example. A crocodile will sunbathe with its mouth open and the thin lining of its mouth absorbs and gives off heat to regulate its body temperature. In the Namaqua chameleon, a more refined way of controlling body temperature through its skin has evolved."

Joseph was handed a mug of tea.

"You see, depending on surrounding colours and heat reflected onto the chameleon by bushes or shadows, sand or grass, so the chameleon chooses to absorb or reflect light and heat off its body by changing its own skin colour. You know, like wearing a white shirt playing tennis or cricket on a hot day keeps the body cool. The multi-coloured skin of the chameleon enables it to maintain a range of cooler temperatures across its body while the environment is conducting and reflecting a whole gamut of light and temperatures."

"That is quite amazing," Mr Salem mused. "Did these results excite Professor Harris?" Joseph's father could recognize his son's expression, see that his eyes were alive with the vigour of achievement.

"He says it's a scientific breakthrough."

Mr Salem brooded for a while.

"Tell me why you do this anyway, Joseph? The relevance of it, I mean. How's it ever going to help me or help you?"

"It's nature, Dad. A thing to marvel over."

Mr Salem smiled cynically.

Jessica detected a subtle change in her father's enthusiasm for the subject. Looking at him, with a degree of worry, then glancing at Joseph, she raked her hair. She knew that what often happened is that her father would begin to criticize the validity or the economics of Joseph's interests, dispelling this biological research as a hobby with no financial gain. It helped to transfer favour onto the farm inheritance instead. As if *Salemkop* wine farm rather than science was a proper job, a proper career for his son. Yet being a diehard nature lover himself with a genuine interest in listening to Joseph's exploits, Mr Salem never noticed how hypocritical he really was in wanting to hear about nature, but then alternately belittling it.

"Is that all you did then?" he asked.

"The work was hard, Dad. As difficult as harvesting grapes!"

"H-mmm..." Mr Salem's tone was unconvinced. His eyes became preoccupied, distracted in following the meandering flight of a seagull beyond the porch windows as it rode the onshore breeze.

"Dad, look I understand how strongly you feel about *Salemkop*. And the farm inheritance." Joseph faced his father's bearded and thickset profile. "Please just give me this weekend to decide alone."

"I hope that now you're a graduate, son, you'll cease the silly beach collecting of animals around Hermanus you're so fond of, and your surveys up the mountains. Especially this weekend of your 21st. All this playing around with nature is part of your childhood. And especially not tomorrow, Joseph, you cannot arrive at your twenty first birthday party covered in dust and reeking of the *veldt*."

"Actually, there are several observations I must complete, Dad, but they are behavioural studies that I can conduct in my bedroom. So don't worry. You won't see me. Nor will Wouter Thyssen."

Mr Salem lightheartedly tugged at his son shirt. As if there was still time for Joseph to throw off the alleged childishness of these hobbies.

"Lexa will be here tonight," Joseph continued.

"Good! I like that girl. You should relax more this weekend, son, take some time off to set your thoughts straight. Be with us a little bit. After all this is the weekend. And we haven't seen you in a fortnight."

Jessica stood and quietly left the men in the enclosed porch lit with radiant sunshine.

*

"Mom, I'm worried about Joe. He wasn't his usual self when he got home yesterday." Jessica put an arm around her mother's waist, watching the woman supervising the maid in the polishing of brassware laid out in a row on the lounge mantelpiece.

Mrs Salem sighed wearily.

"It's going to be a difficult decision for him to come to terms with, Jess." She looked at her daughter with sober concentration in her eyes. "Having to give up his biology studies."

Hearing this conclusion, this apparent career eventuality for Joseph, hadn't been voiced so clearly before by Joseph's mother. This surprisingly blunt response of hers startled even Mrs Salem. It was the strict finality of it. For she wasn't a woman prone to bluntness or assertiveness, nor any sudden trenchant decision. But as Joseph's birthday approached, the stark reality was the inheritance situation. She could see no other avenue for her son. Joseph would have to capitulate. She had now resigned herself to this eventuality, knowing too well what her son was up against in his father. Simultaneously she tried inwardly to comfort herself in the knowledge that Joseph couldn't anyway be a student his whole life. In this aspect of motherly caring, she was no longer sure whether her role as a mother was simply to wish happiness and

67

protection on her adult children. Because her husband already faced this reality of them being adults.

"I heard Joe cursing in his room last night. He threw something against the wall."

"No. Surely he couldn't have!" Mrs Salem's exclamation held a mixture of alarm and disbelief. "He seems alright to me. Just a little quieter perhaps, but he isn't sombre or depressed, is he? You know how Joseph gets preoccupied with his own thoughts."

"Well, he wouldn't even talk to me last night," Jessica challenged. "He wouldn't even unlock his bedroom door when I went upstairs and knocked on it."

Mrs Salem examined the polished finish of a brass ornament, taking on an expression of caring, then turning in Joseph's defence.

"Jess, obviously he's got a lot on his mind. When he comes to realize that this decision is also for family, the whole family, Uncle Jack's side too, I think everything will resolve itself."

Suddenly Jessica felt alone. She dropped her hand from her mother's waist and paced across the room towards the piano. Wondering who's side was her mother on anyway? How could her mother preempt Joseph's feelings? She of all people? How could his love for biology be cast aside so easily, when he obviously hadn't yet made his own choice.

She sat down at the piano, desperate to pound out the truth with her long and beautiful fingers. The fingers of a born pianist. Thinking that a loud concerto by Rachmaninov would suit the occasion. But instead, before her, was the score of the theme from *Elvira Madigan:* second movement andante, piano concerto number 21 in C major, one of Mozart's loveliest dream world slow movements. Gently she struck the first notes. Sad. Sensitive. Music had become entwined into Jessica's moods, it mirrored them, and they reflected strongly onto each note. Soon she became lost in the heights of Mozart.

"I WILL NOT HAVE THIS INDECISION ABOUT YOUR FUTURE!" Mr Salem was heard at the top of his voice from the porch. In the background Joseph seemed to be trying to passify his father not to get this overwrought and excited.

"But Dad, but..." he pleaded.

Before Mr Salem bellowed: "NOW THAT IS THE LAST WORD I WANT TO HEAR FROM YOU!"

Then the conversation reached an abrupt end.

Jessica plummeted down to reality and stopped her rendition of Mozart. She lifted her fingers from the ivory keys of the piano and raked them through her hair, looking up to see her mother's worried expression as Joseph passed through the lounge, walking quickly. He exited immediately and took the stairs up to his bedroom.

Together they noticed the complexion of Joseph's face: it was drained, pale white as a sheet.

- CHAPTER 3 -

Joseph trod the steps to his bedroom, counting them out of habit, there were twenty two in all. He reached the upstairs landing and shouldered aside his half-open bedroom door whilst straightening the jackal skull nailed to it. To him there was nothing macabre or odd about this ornament or, for that matter, any other of the objects cluttering his room. Each piece of natural history had been collected carefully, thoughtfully, for study, and every one told an important story. Mindful of conservation and assured not to disrupt ecosystems, these animal and plant specimens had been collected to preserve a record from road kills, the spoils of illegal poaching, surveys of biodiversity, rare plants saved from damage by human development or *fynbos* fires, and marine life recovered before the tourist gaming or diving seasons.

Joseph crossed the polished oak floor, taking a seat on the bed, his back leaning against the paneled wall to face the row of sea windows. A blue-pointer Siamese cat named Mowgli left its resting place on a windowsill, strode across the distance, and jumped up purring to nestle on his lap. The cat was accommodated with a gentle stroking as it nudged Joseph's knee.

Perhaps Dad has been right all this time, Joseph contemplated. A career in science is hardly a money spinner. Science is hard work that few others understand. It was a sober thought.

In the cascade of morning brightness radiating from the roof windows it seemed the wrong time for such seriousness, such sobriety. But it was true nonetheless. Science as a profession is a tenuous, difficult struggle: for funding, peer recognition, for elusive scientific results that in the end - after a life's work - may just as easily be discarded by society as unimportant or of little use. If his father's judgement and opinions were the deciding factors, his interests in science should remain a hobby. Professional scientists can become laboratory or study bound, often working antisocial hours, living from one conference to the next. How did that fit with Salem family responsibilities? Field biologists are worse off still since researching nature in the field attracts very little financial reward.

The alternative was *Salemkop* wine farm. The family inheritance. And his father's ambition for him to become farm manager with Uncle Jack. With the absolute certainty his father expressed in the belief that Joseph would succeed in this venture. Easily succeed. Enough fatherly insistence had eventually forced Joseph to begin to view his passion for science differently: to attach a heightened sense of value to these farm securities. Or were they insecurities? Were they, in fact, his father's insecurities? What could not be ruled out as an influence, Joseph knew, as cause of his father's insecurity, was his tragic accident. Before the accident, his father always dominated his brother Jack. Never trusted him. Even though they were joint owners. His father always had to be in control. The accident, however, had forced him to relinquish control of the farm to his brother. Whereupon his distrust of Jack had become an obsession. Does Dad expect me to wrest back control from Uncle Jack, Joseph wondered?

From the bed Joseph glanced at his seated reflection in a far wall mirror. His face appeared pale, even gaunt. He still felt shaken from that last blow-up with his father downstairs. From this angle his face looked drained, never mind how his skin was bronzed by the sun and framed by the curly darkness of his hair. The mirror made him appear deceptively long-legged on the bed, almost gangly, his six-foot one inch height seated implausibly with crossed legs. Perpetual outdoor activities had produced a slender, taut physique, firmly muscled. He listened to the force of the onshore wind striking the panes of roof windows with a humming sound, the rumble of breakers muffled by the

73

double-glazing as waves advanced endlessly in formation upon the shore.

So many objects in his bedroom told a story, even these windows. After Mr Salem's accident and before permanently moving to Hermanus from the farm three years ago, the family enlarged this holiday cottage. Through Buck's and Joseph's insistence, *Rus en Vrede* was transformed into a genuine double-storey: to them it was a laughable pretension that the house previously had roof windows merely for show. Especially since every holiday the two of them were employed to clean the salt-caked panes. Each year they balanced precariously on dusty beams inside the unconverted roof, scraping off encrustations of salt, polishing the glass, replacing net curtains perished in the sun. In the process of doing this, Buck and Joseph joked secretly among themselves that like these false windows, many things in South Africa were also false.

"To overcome falsity," Buck had challenged, "it should be measured for some practical use."

In the end these roof windows, objects of such risible contempt those years ago, did become functional. As the row of sea-facing windows in Joseph's bedroom now spattered with a fine sea spray. Jessica's bedroom remained downstairs converted from the shared bedroom they had both occupied as children on holidays, and the porch gained an extension and was glassed-in. During his four years at university Joseph's loft bedroom transformed into a sleep-in laboratory: family visitors expressed some considerable amusement as to his taste in bedroom lifestyle and furniture. And in reply he would laugh with them, in equal measure seeing this weird side of his personality, while at other times allowing their deep misunderstanding of his passion for biology to simply ride.

Buck had used his fishermen's carpentry skills to build the floor-to-ceiling bookshelves along the west wall which Joseph now faced sitting on his bed. Buck had also helped fit the L-shaped dissecting bench with basin, and the work-desk with optical microscope. Two metal filing cabinets purchased in Cape Town had been transported, to the entertainment of the coach passengers, all the way home in the bus with Joseph. Timber crossbeams intersected the slanting ceiling off which hung an assemblage of objects: a stuffed specimen of a southern albatross with its superb open wingspan; two model pterodactyls from his childhood days, one of these prehlstoric repliles in a steep dive near the bed; while a bird cage hanging by the rear window had a live occupant inside. University awards and prizes of money had allowed Joseph to install some of this furniture and to buy a computer himself.

Arranged on the dissecting bench were four baboon skulls (all from the same troop) bleached white by the sun; the complete constructed skeletons of a Cape cormorant and a Cape rock rabbit, and the sequence of spine bones of a Cape fur seal which had been recovered from the stomach of a great white shark. In the myriad of bottles filled with formalin preservative were the foetuses of a mongoose and of a pygmy shrew and many other items; there were fern and reptile fossils collected from a Karroo riverbed; jars of sea-cucumbers, larvae of fish and crustaceans, and many types of seaweed. A hand-shaped piece of orange coral was pinned onto a bookshelf near a giant tropical conch shell with spines, and there was part of a dry old termite mound cut in cross-section. The wall displays contained local butterflies and moths collected in and around *Kwaaiwater*. While hidden in the natural history drawers were a complete catalogue of local bird eggs, pressed herbarium specimens of Cape flora, rock and shell collections, and rodent study skins.

But if Joseph were to choose the most precious showpiece of all, undoubtedly it would be the first article he had ever published in science. It was framed behind his bed. This scientific article went by the title:

'A new Southern African whip scorpion.'
by J.W. Salem and E.M. Harris.

This remarkable achievement for an undergraduate student was Joseph's first formal contribution to science. He had stumbled upon the unique species of whip scorpion, a strange pincered animal with a whip-like tail which secretes an irritating vapour, during a biology field camp and was urged by Professor Harris to publish the discovery. Immediately he had set about the task. First preserving the original type specimen for posterity and reference, notifying important museums, choosing a Latin name for it, and then describing its unique features in scientific detail. Once published, Joseph's name would forever be associated with this whip scorpion.

When the article first appeared in print a year ago, excitedly he had rushed back home to Hermanus from university in Cape Town. Rallying the family together, he had explained its contents: his mother liked seeing his name in the title, Jessica commented on the fine artworks inside, and although his father commenced reading the article with great enthusiasm, Mr Salem afterwards became quiet and withdrawn, assuring everyone that he needed no further explanation of the subject.

This morning, lowering his gaze from the wall mirror, Joseph leaned over to the bedside table to retrieve his diary. Paging through he reached his last diary entry dated 21st November to 5th December 1986:

Namaqua-chameleon fortnight.

Week 1 in Namibia we spent in a true sandy
desert with midday temperatures soaring above
40 degrees Celsius.

Hot enough to blister the skin without block-out sun protection. The windblown sands sting your skin like insect bites. In the silence of the desert, distant mirages shimmer and glisten.

Day one: Professor Harris and I visited the *Welwitschia* flats where the wonder plant of the Namib desert, *Welwitschia mirabilis*, grows. It is one of the world's longest-living plants (each plant survives hundreds of years) and endowed with all manner of legends. Even Dr. Friedrich Welwitsch who discovered it in 1859 (in the same year that Charles Darwin published 'On the Origin of Species') was hesitant at first to touch a specimen because he feared it may be an illusion.

They are specialized living fossils possessing features of both cone-bearing and flowering plants, but are unrelated to any existing flora.

[See: taxonomic specimen ND1742 which Includes a sample of beetles taken from a cone of a female plant]

Later, back near the research station I walked alone along the Kuiseb riverbed, finding various diggings where larger animals had tried to reach underground water but failed. The skin and bones of a gemsbok carcass lay closeby dessicated in the sun and no longer decomposing. That evening game rangers from Etosha National Park arrived to discuss their lion contraception programme...

Joseph skimmed over the aim and methods of his chameleon experiments, through the long list of scientific equipment used for his trip: thermocouples, portable

78

photometers, etc, searching for the description he wanted of that second week. He read:

Late on Wednesday afternoon I took the
Computer analysis of my chameleon project to
Prof. Harris.

As Joseph gazed out again beyond the row of sea windows, placing the diary onto his lap, he recalled how intellectually inspiring and uplifting were the discussions with Professor Harris when travelling across Namibia. It was the first time he and the Professor had conducted a field expedition alone without a whole class of undergraduate students accompanying them. Initially Joseph was nervous to be alone with the professor. What would they speak about? What about embarrassing silences? His fears proved to be unfounded. For the first time Joseph felt treated as an equal, as a science colleague rather than a student. Professor Harris even once offered fatherly advice, advice which was unlike Joseph's own father where a respect for his talents as a scientist was concerned.

A week later, back in the Zoology Department at university, he recalled the informal way he had knocked on the door of Professor Harris's office and entered. Relaxed, sitting in an easy chair with his feet casually put up, the professor remarked how he had looked forward to their meeting the whole day. Carrying a thick wad of computer printout under his arm, Joseph unfolded it onto the coffee table beside the professor's legs, and together they put their minds to it, reams of data and statistics, and mulled over the experimental results.

"Brilliant," he said finally, nodding to Joseph. "I think we ought to publish this work." Professor Harris smiled at Joseph's surprised face, stood up, shook his hand. "But this time I think we should aim for an international science journal," he said with authority.

Even more unexpected is what followed. The professor crossed over to his desk and extracted two official university envelopes from a drawer. He handed them to Joseph. The manilla envelopes were addressed 'J.W. Salem Esquire'. The first, which Joseph tentatively opened with a straightened forefinger, contained the offer of a university place next year as a doctoral student. The second was a Ph.D. bursary covering his forthcoming study fees for the doctorate and living expenses.

"This doctoral degree," Professor Harris explained, "will be the basis of your specialist training as a research scientist. Very few students are afforded the opportunity, Joseph. So I need you to think seriously about this and come back to me with your decision promptly."

Leaning back now against his bedroom wall, still astonished by the momentous offer of a doctoral place, Joseph began to consider his reactions to it on

Wednesday and yesterday while he stroked Mowgli. The cat purred in loud contentment.

Following that Wednesday appointment, Joseph had left Professor Harris's office with what felt like a fire in his veins. As if he was suffused with a double dose of life: mainlining on dreams he had entertained since his childhood, entertaining the deepest, purest visions of his future. Through supper in the university canteen, Tim, a fellow student who shared an adjacent room in residence, had nudged Joseph.

"You haven't stopped smiling for ten minutes! I give up, man? What have you been smoking? And damn it, why are you so impatient tonight, quit your incessant foot-tapping at the table!"

After dinner, Joseph had phoned Lexa with this news. But later that evening while packing his bag for the weekend trip back to *Kwaaiwater*, another thought struck him. A darker thought. A thought with a sombre undertone. How do I explain this to Dad? Suddenly too many demands were converging in his brain at once: his 21st birthday looming this weekend, the farm inheritance, this new doctoral offer that urgently needed an answer. In the conflict of emotions, alone, isolated, suffering behind the closed door of his room in residence, Joseph collapsed on the floor with an attack of agonizing stomach cramps. Twice that night he awoke in a shivering sweat. He gulped down antacid stomach pills which he located in a medicine cabinet at the end of the hall. For a few hours the pills helped to quell the rippling spasms of pain in his abdomen.

On Thursday morning Joseph woke with a pounding migraine headache. The pain was debilitating. A splitting and throbbing in his head. It was hard to wrench himself out of bed. This sapped energy, this paralysis in what to do, the uncertainty about his future, was so unlike his optimistic usually driven self. To quell the headache he plunged his aching head into a basin of ice cold water. By late morning, when he finally emerged from his residence room, he decided he couldn't be with people, and crossed the university campus to spend some quality time in the library alone. After lunch *en route* to catch the mid-afternoon coach to Hermanus, he randomly bumped into Professor Harris. Joseph's first ambivalent reaction was to quickly turn away from the professor and not acknowledge him. It was a stupid regrettable reaction, he knew, which afterwards left both men bewildered.

From this vantage point sitting on his bed in his laboratory bedroom at *Kwaaiwater*, far removed from the reaches of both the university and his father, Joseph could begin to contemplate more soberly a solution. So what now, he thought. Find a way to talk with Dad? Persuade him to listen to Professor Harris's offer? Realizing that he was slowly pounding his fist against the bed, Joseph stopped. The situation seemed futile. Hopeless. How often and for how long had he tried to reason with his father. He doubted trying to understand what motivated his father, his father's insecurities, obsessions, would help to make a difference in changing his father's mind.

Joseph remembered how in the trauma following the accident, his father had rediscovered the bible which had helped him to overcome the momentous change in his lifestyle. A new life of religious belief. To match his life of being wheelchair-bound, of trying to overcome his disability. The loss of his career. The loss of his status. In this context Joseph felt it impossible to criticize his father's needs. Or the desire his father voiced that the farm inheritance pass to Joseph. But as his son he had to consider an event the family had long ceased discussing. A subject now forbidden in the *Kwaaiwater* house, a subject which was considered a disaster in Salem family history.

That event was his father's accident...

*

A mere three years had elapsed since the accident. It seemed to Joseph a lifetime ago, yet it occurred only a year after Joseph entered university. The family lived at the farm at *Salemkop*, Mr Salem was joint manager with Uncle Jack, fit and burly, full of the vigours of farming and his wine-making achievements. The inheritance expectations placed on Joseph were entirely different. Pressures on him to follow family tradition were less urgent, flexible, although it was made clear his university education should include botanical courses in viticulture, to learn the technical side of growing grapevines. Grateful for the financial support given to start his science degree, Joseph had complied.

It was Easter at *Kwaaiwater* on the day of the accident. When the Cape autumn is settling in. When from one day to the next there is an apprehension among locals whether the Cape of Storms will live up to its name. A storm had recently rumbled by with menacing force and an unpredictable offshore wind rattled the coastal vegetation. With the tide high, the sea in *Voelklip* cove was calming though still choppy and murky. Perfect conditions for fishing, they guessed. So Mr Salem and Joseph equipped themselves with rods, tackle and bait, and crossed the rocky tidal terraces to the water's edge. In the space of three hours they landed two sizeable *galjoen* and an above average size white *steenbras*.

"I think that's enough!" said Joseph loudly after gaffing the last fish. "We should call it a day, Dad." These precise words were etched into Joseph's mind. "Let's head back now. The horizon is building with cloud. The wind is picking up." But his stubborn father wanted to equal the number of fish caught to two white *steenbras* also.

"Go if you want to son," Mr Salem shouted against the wind. "I'll be half an hour tops."

And so Joseph made the decision he always regretted. A decision which still haunted his dreams. He left his father to continue fishing. Forty minutes later, the family began to worry. Joseph raced back feeling suspicious. The first thing noticed, he remembered distinctly, were the rocks. They were now wet and slippery. Dark brown with wetness. Then he located the overturned tackle-box and the gaff pointing skywards lodged in a crevice. The fishing rod and his father were gone.

The wave we've always talked about, Joseph uttered under his breath! He cursed himself in sudden panic. He began running towards the rocky edge tearing off his outer clothing, woollen jumper, sweater, track suit bottoms, while the shells of sharp barnacles underfoot slashed at his shoes. His sight became distorted by a quickening heartbeat. Schloooop... hissss, schloooop... hissss was the strong suction sound under the overhang of rocks as the heightened ocean swell advanced then retreated. By the time he reached the overhang edge, Joseph was soaked in spray, blinking furiously, and shaking his dripping hair to clear his vision. Then he lay flat on the rocky overhang to peer under. Far beneath, on an encrusted ledge, his father was clinging like a limpet.

The man, in his stubborness, was desperately hanging onto the rock-face but being submerged at each swell. In the vacuum of air before each successive swell, Mr Salem could be heard gasping loudly for breath. At the next outsurge Joseph dropped into the water wearing only shorts and a T-shirt. He shivered at the icy power of the ocean. Swimming with the swell, it carried him under the overhang, his arm gashed by a jagged outcrop of rock. Caught between the swift rising waters and the roof of curved rocks he dived and reached his father. But the man, rooted in his own survival instinct, would not let go.

Joseph tore at his father's fingers yet they just firmed and tightened around the rock. Punching at his knuckles and with an arm wrapped firmly around his father's shoulder and upper torso, Joseph pushed resolutely away from the rocks with both legs. However, it was the next outsurge that did it. It synchronized with Joseph's immense effort and swirling together they were carried out on a rip current into the cove.

His father became limp: moaning that he could not move his legs. By some miracle he was still breathing although semi-conscious, coughing, spluttering, murmuring incoherently. It took all of Joseph's energy, all his strength to keep the dead weight of his father afloat. In turn Joseph found himself gulping salt water in hiccups while trying to hold the man's head high above the choppiness of the sea. After several minutes the rip current washed them to its end and deposited them in calmer, deep water. Joseph treaded water laterally, steering into the shoreward swell and slowly towards the beach.

"That you son?" Mr Salem croaked through blue lips. "Is that you?"

"I'm here... It's me, Dad. You'll be alright."

"Please God, don't leave me Joseph, don't ever leave me! Mr Salem repeated these words in a state of delirium, gargling and spitting water. Apart from the blue coldness of his skin, he was trembling badly, his lacerated hands clinging to Joseph's shoulders and his hindlimbs trailing behind as limp and flowing as the tentacles of a sea anemone. Joseph's arm was seeping a trickle of blood that as rapidly as it seeped and streamed was diluted in the water. Mr Salem was passing out intermittently, his jaw held firm in his son's grip.

On the distant beach Joseph could make out a cluster of figures: his mother recognizable holding her hands to her terrified face; Jessica laying blankets out on the sand; there were neighbours; and Buck's dog barking at the surf... and, yes, there was a *coloured* man racing and diving through the waves. Buck swimming out here! Joseph sighed thankfully.

When Buck arrived on the scene Joseph was near to his limits, softened with fatigue, and his muscles ached. As they reached the wave-zone, Mr Salem started to scream in pain. The man's face screwed up in agony, no matter the intense cold of the water which had anaesthetized all three of them. It was then that Buck and Joseph sensed serious trouble. The churning wave-action and undertow toyed with Mr Salem's paralysed legs as though his limbs were made of rubber.

*

Taking the accident into account, considering the enormity of what happened on that unforgettable autumn day out in *Voelklip* cove, Joseph realized that anyone, almost anyone under such circumstances, could have turned strongly to religion. Why not? It was close to a miracle his father survived. During the five months of hospitalization and rehabilitation, Mr Salem found a great inner strength in the bible. A new religious leaning. The book itself became a companion. A silent saviour.

In those months of regular hospital visits the family grew closer, more cohesive, more understanding. It was a time when Mr Salem discussed his chances of remaining a wine farmer. Medical opinion confirmed the likelihood he would be confined to a wheelchair for life. *Salemkop* farm was no place for a disabled man: uneven gravel roads, high steps up to the farmhouse, steeply inclined, uneven land. As compulsive and active a farmer as Mr Salem had been, it would not only be unwise and unfair, but exceptionally frustrating indeed for him to continue as manager. Unable to supervise his labourers from a wheelchair, never mind the irritation of being unable to attend to manual tasks which he himself relished. The family therefore decided on a permanent move to

Kwaaiwater. They would renovate their seaside holiday home.

Personal relationships underwent great flux too. Mr Salem's style of communication with family altered dramatically, which Joseph especially noticed. For four of the five months recuperating in hospital, Mr Salem's predicament and increased religious leaning humbled him beyond recognition. In the man's eyes, of course, there remained a burning desire to work again, to get outdoors, to be a farmer. But it hardly consumed his caring for others. Some remarked that the accident had turned Mr Salem meek. It came from people in the farming community, those who knew him less, those less tolerant who admired Mr Salem for his bombastic, straight-talking shoot-from-the-hip attitude. For his no-nonsense strong man stance as a respected *white* farmer.

Yet Joseph rather liked to think that in this newfound humility, his father had found a greater inner strength. It was a belief which proved itself when he was in hospital in the countless new hospital friends Mr Salem attracted to his bedside. For every new receptive face he offered a proverb: a proverb to compliment the occasion, a proverb that left the recipient with a slice of wisdom to take back to bed, a proverb that brightened the place with hope. Optimism. Thoughtfulness. Love. The Salem family in equal measure were both surprised by this transformation and warmed by this new profusion of caring. It ignited an added closeness between them that mirrored the affection other patients were showing towards Mr Salem. It was as if his new sensitivity, new awareness of others, stemmed from the experience of great pain.

But that fifth month of rehabilitation was a mystery. An unexplained period of reversal, regression, seemed to take place when the counselling ended, after which Mr Salem began to alienate himself from his newfound friends. To the extent he interpreted the proverbs to his own end. What exactly happened? Joseph had racked his brain on this matter umpteen times. The doctors suggested Mr Salem was suffering "regressive anger". That his father was realizing, in fear and rage, the enormity of accepting he may be paralyzed for life. It happened in the month he was to be discharged from hospital. Only then did body scans and blood tests confirm that Mr Salem had suffered a mild heart attack during the rescue when he was being pulled through the surf, irrepairably weakening his heart in the stress. It was the same month in which renovations on *Rus en Vrede* were completed and Mr Salem felt self-assured he might even be able to run the farm from the distance of *Kwaaiwater*. Around this time, he imposed his searching question onto Joseph:

"Why didn't you wait for me while I was fishing, son?"

"Excuse me!" Joseph responded. "What do you mean!" The question seemed so angry, so selfish, so unfair. Patently his father's recall of the event was untrue.

They were imposed words, intolerant words of impact, which had the effect of freezing Joseph to his core.

In her conciliatory way, Mrs Salem gently tried to explain that it was no one's fault. In his own defence, Joseph repeatedly pointed out that he had returned home with his father's permission, indeed his blessing, because Joseph had been expecting a phone call from Lexa. Yet Mr Salem admitted none of this. He would acknowledge no conversation whatsoever with Joseph on those rocks when fishing. The hospital physicians warned the family of the psychological effects of the trauma: Mr Salem was on medication for it, and that if there was love, patience and understanding, there was no reason why he should not adapt at home.

The final blow struck Mr Salem when doctors urged him to retire. His heart complaint was serious enough to warn of dire consequences were he to resume farming. It was after Mrs Salem worriedly explained to the doctors the sometimes difficult clash of personalities between her husband and Uncle Jack. She was panic-stricken at the thought anything else might befall him, triggered by these other family stresses. And eventually Mr Salem recognized that his wife too had suffered enough. He had little choice. He even stopped smoking cigarettes for her sake.

"I'll only smoke a pipe now," he relented.

*

Where had this brought Joseph today? He couldn't help wondering, recalling this long chain of events, how far his father had really developed into that four month dream man of his hospital days. Sometimes the family felt excited, an exhilaration even, at each small step in his progress. But at other times they were rudely brought down to earth. However, there was no doubt about family camaraderie, about the family's persistent hope that it would all resolve itself in the end. Joseph learned to tolerate his father's use of his heart ailment to gain sympathy and especially to dominate or end a conversation. Working together as a team the family silently understood not to fluster or overstress the man.

"But what about my life!" Joseph silently muttered to himself on the bed, clenching his fists. His movement woke the cat and Mowgli jumped off the bed. At a time like this, when he needed to discuss with another sober and rational mind the amazingly kind offer of the doctoral research position, his father asserted such a powerful influence. "Damn it, damn it!" Joseph cursed at the manipulative hold his father wielded. In being disabled, how he exploited sitting in a wheelchair, used his paralysis to exempt himself from other people's points of view. And the sheepish, timid way in which the family had learned to rally to his father's favour.

But now Joseph felt an imperative. Responsibility. Pressure. Deciding to take up the doctoral position had a deadline that must not be missed. Which coincided with the decision whether or not to sign the inheritance at Salemkop on Monday. Joseph bit his lip while his fists clenched until the knuckles turned white.

"Damn it! There comes a time to reject certain family expectations. This weekend I will have to confront Dad. I must confront him... Ahh!" A corkscrew of pain twisted itself in his stomach as a cramp which shot through Joseph's abdomen. It felt like the pain on that night of agonizing cramps he had experienced in the university residence.

Joseph rose slowly from the bed. Holding onto his stomach in pain he walked over to the adjoining bathroom and in one gulp swallowed two painkillers with water. He was shocked at the appearance of his face in the mirror: grey complexion, frown etched on his forehead, darkening rings around his eyes.

"I've got to get some fresh air", Joseph contemplated inwardly. Perhaps Buck is still down on *Kwaaiwater* beach throwing dried seaweed for his dog to retrieve. I need his advice. I need to talk to him about this. He will be wise to much of it. "Yes", Joseph brooded, "and I've got to get out of this house!"

- CHAPTER 4 -

A boisterous onshore wind struck Joseph on the side of the face as he exited the front door, and by the time he crossed the boundary fence of *Rus en Vrede,* his hair was disheveled, gleaming with auburn streaks. But he felt free of the cloistered atmosphere of the house. More especially he felt he had escaped the ambience on the porch. As the wind blew hard, as he absorbed some of its brute force, its fresh salty smell, he could feel the very fibre of his being, restored. So much of the beauty of *Kwaaiwater* relied upon this wind that sheared the vegetation, though now little could be heard of other sounds above the sweeping, rattling tumult of the wind in the leaves. Joseph stopped to listen but the Labrador barking had long ceased.

Guessing that the old fisherman may still be on the beach, his footsteps quickened down the steepness of the coastal path. The small-leafed vegetation around him was thick and dense and it suddenly blocked out both the view of the beach and the houses; it was like walking through a manicured maze with blind turns of hedges following the contour of the land. Of the two coves at *Kwaaiwater,* this path traversed the smaller cove enveloped in bush and sculpted by the waves into a semi-circle of high rocks. In the smaller cove the swimming was safer, its beach sand smooth and white. This is where Mr Salem on that Easter day was dragged ashore.

Joseph entered under a canopy of vegetation, a bushy tunnel that was warm and extensive with a mat of twigs and dry leaves on the ground that exuded a pot-pourri of natural lemon odours. In the wetter shadowy places, white arum lilies were growing in leafy clusters. Then the path opened again into bright sunlight steepening up to the promontory that separated the small cove from the main beach. The gale was strong enough to be shredding furrows through the thicket as he reached *Kafferklip* lookout. Gazing down onto the main beach, Joseph held his ears with cupped hands against the wind. His cupped hands hummed. From this vantage point it felt as if here the whole town of Hermanus fell into insignificance, swallowed up in the breadth of Walker Bay.

He could see no man nor dog. Only Annie and Joyce were on the beach busy collecting their shells as usual. Joyce lived next to Buck in the *Coloured* quarter, so she would have recognized that whistled tune: '*I was born under a wandering star*' or the whine of Buck's harmonica

95

or the barking of his dog. He rounded one more bend before the thicket cleared as the path led onto beach sand.

The wind was subdued down on the beach. As Joseph traversed the coarse sand it crunched noisily underfoot. The copper toned water of *Mosselrivier* glistened quietly as a slow flowing stream and he crossed it at the little wooden bridge, glancing for a moment at the mineral waters that trickled away into the sea sand. He negotiated his way towards Annie and Joyce over piles of shells of black mussels washed high up onto the beach by the tide.

"Hello Joyce, how goes it?"

"Fine Master Joseph." The two elderly women had noticed his approach from the bridge and Joyce now stood up. "Weez almost collected enough mussel shells for today: the purple and black ones."

"Is Mr. Van Rooyen still taking them?"

"Ja."

"Is he paying you enough?"

She shrugged unsure. "I think so." She nodded. "Because he gives us the small coloured bottles for free. Then we pour our polished shells into them as an ornament. Mr. Van Rooyen says he gives us a quarter of the profits. Last week - its true Annie, isn't it - he sold three dozen of the blue type of ornaments. He says it's due to the holiday season!" She explained this with a clear enthusiasm in her voice, smiling through toothless gums.

"Have you seen Buck today, Joyce?"

"Nee, Master Joseph," she said, shaking her head. But then turned to her fellow shell-collector, asking the other woman rapidly in Afrikaans. With the added information given to her by Annie, she changed her mind and nodded: "Ja. Annie heard him about ten o'clock. I reckon he's at home, Master Joseph."

Joseph put a friendly hand on her shoulder in appreciation of this information. With his touch, Joyce's face lowered suddenly as she giggled with embarrassment at this familiarity and shrank an inch away from his hand. Unlike the boldness and independence which Buck displayed, Joyce, and Annie herself too, harboured frightened souls low in self-esteem. This was typical of many of the older *Coloureds* in the town who had lived under forty years of apartheid, experienced extreme poverty, notwithstanding the hardships associated with being a woman. Forty years of racial segregation. A whole lifetime of enforced discrimination. Whose sensitive natures shrank from the white man. Joseph had asked Joyce and Annie many times to avoid using the prefix 'Master' when addressing him, but they remained unsure and uncomfortable changing this.

For a brief interval before departing, Joseph watched as their hands sorted through mussel shells. Small leathery hands cracked from sea salt and nicked from the razor sharpness of the shells. On his route towards the main tarred road, Joseph contemplated the short-cut he could no longer take because of his father's pact with Wouter Thyssen. Passing by the *white's only* hotel that overlooked the cove, on the face of it there were no outward signs of the country's State of Emergency. Why was his father so worried? Of course, petty apartheid abounded throughout the town: separate public toilets for *whites* and *non-white*, a *white's only* cinema, even the beach was barred to '*non-white*' leisure activities, but none of this represented a change to the face of Hermanus.

Like all who belonged to the privileged, these prejudices at first hardly affected Joseph nor did he really notice them. As a boy he made friends with whom he pleased, moved around as he liked, but by his mid-teens when friendships were established with some of the Hermanus *coloured* folk, this apartheid politics blatantly got in the way. These friends couldn't share public places with him. Rumours began to circulate about Joseph's loyalties. Words trickled to Mr Salem. When these rumours became an embarrassment to the Salem family, Joseph's father stepped in to prevent these *coloured* friends from visiting his son at home. Unless, of course, it involved employing these friends in useful labouring jobs like gardening, house painting, desalting the loft windows.

Isolated in a poorer corner of the town was Hermanus *Coloured* Quarter. Standing at its entrance, Joseph could distinguish Buck's house far up the gravel road. The door appeared open. He strode down the pot-holed road. On the muddy pavements strewn with garbage, children were playing happily in the mud and the dirty water. Mongrel dogs sniffed around the scraps. Joseph hopped across stagnant puddles, breathing in less deeply so as not to inhale the stench of rotting offal. Most tenants of these two-roomed brightly painted houses, all the way up to Buck's place, Joseph knew by first name. As he passed by, some residents waved to him in acknowledgement. Katrina who saw him approaching up the road called out in her strongly guttural accent:

"Master Josef, Master Josef!" She withdrew into her shack and appeared again from the dark interior with a jar in hand. "Hullo Master. Here is for you." She handed Joseph a handsome specimen of a scorpion pickled in dark vinegar. Turning the jar slowly around at eye level, he examined it carefully.

"Where did you find this, Katrina?"

"In the toilet last week." She pointed to a corrugated tin outhouse shared by five of the surrounding families. "Jannie, my eldest boy, nearly stood on it. Now we always lift up our feet when we sit on the toilet." Joseph delved into his pocket and extracted two fifty cent coins.

"Here's fifty cents for the scorpion, Katrina. And fifty cents for the vinegar. Thanks."

She curtsied politely. "You must come for tea sometime."

"Next time I'll visit you properly. Also, I'll bring a present for Jannie." Joseph waved a hand at her younger son, Cornelius, who was sitting dirty in the vegetable patch. "Can I offer you a bit of advice Katrina, if you find another scorpion?"

"Ja."

"Be careful of the dark grey scorpions that have small pincers but a thick tail, like this one." Joseph demonstrated the action of the pincers by opposing the forefinger and thumb of his right hand repeatedly. "Grey scorpions are much more poisonous than the smaller orange scorpions which have a thin tail and big pincers."

She nodded when he pointed out this anatomy of tail and pincers from the pickled specimen in the jar. After explaining what first aid could be given in an emergency if any of her family was stung by a scorpion, Joseph departed and continued towards Buck's ramshackle house tucking the specimen jar into his hip bag.

*

"Come..." a gruff voice answered as Joseph knocked on the half-open door. The voice sounded educated, it seemed out of place in this squalor. "Ahhh, *muishond-kind!*" Buck exclaimed. "How goes it *seuntjie*. Long time no see... and why the sad eyes, why the frown?"

The old fisherman was sitting in the cross-hatched light of shadows thrown by a fishing net. Like merlin reclining in a cool damp cave his curly grey hair and beard and his lined face gave him the appearance of a great sage with the bluest, azure blue eyes. It was hard to believe he was seventy-two years old. His thickset body was taut and well-muscled with a weathered skin, kept through years of harsh manual labour, invigorated each morning with a swim in the pounding surf.

"Is my expression that obvious?" Joseph asked.

Buck motioned for him to come closer in the mottled light framed by nets drooping off ceiling hooks. Then he stood up and placed both hands on Joseph's shoulders, squeezing tightly, his eyes looking upon this 'mongoose-child'. The name *muishond-kind* had been chosen years ago by the *coloured* folk when, in his teens, Joseph adopted and reared a Cape-grey mongoose from birth.

"Today you're not the happy rock-hopper that I know," he said. "But first, pat Randy over there to show him that you care. His tail has been wagging hell-for-leather ever since you came in."

Joseph obliged and the Labrador licked his hand with a wet pink tongue. Joseph walked on over to a low-slung hammock facing Buck and lowered himself into it in a seated position, rocking slowly on his toes. Glancing at Buck, then at the dog, he began to pick at the frayed canvas lining of the hammock as it squeaked with the rocking motion. For the first time in this place Joseph found himself at a loss for words, frown etched on his face. A prolonged silence commenced between them.

"I can see how difficult it is for you," Buck at last intervened. "Tell me later if you please. How was your trip to Namibia?"

Joseph shrugged. It seemed to him a loaded question. Answering this question would only include Professor Harris and introduce the whole of his predicament.

"You know," said Buck with a reflective voice, "twenty five years ago I worked the trawlers in Walvis Bay close to where you visited. Sardines. Mullet. I was there long before you were born: before those multinational exploiters took over. Hells bells, those were good foggy days in the Benguela before they overfished the industry."

Randy crossed the room and buried his head in Joseph's lap, his tail slowly wagging.

"Are you listening, son?" Buck's tone changed to mild alarm. Joseph was sitting with a somewhat dejected posture, eyes lowered, indifferent, the hammock swinging back and forth with his weight. Buck paused trying to change to another subject that didn't include Namibia. "So what's been happening with your environmental group. Have you organized anything more?"

This subject drew Joseph from his reverie. It seemed neutral to how he was feeling. He reached into his hip bag for a folded sheet of paper and offered it to the old fisherman's outstretched hand. Buck read:

FACT

(An organization committed to exposing environmental and social injustice in South Africa)

AIMS:

1. To research environmental and social problems in South Africa and to relate these to their causes using scientific procedures, history, public opinion, and international law.

2. To publish accounts of present-day environmental and social injustices and to offer predictions for the future. This publication will be distributed to relevant national and international organizations involved in the the protection of natural resources and the abolition of apartheid.

3. FACT is not political, rejects having any ideological base, and will not receive funding or pressure from any political bodies. All articles published by FACT will present well corroborated evidence and their conclusions will highlight the extent of the injustice in relation to its cause.

4. FACT aims to uplift the environmental and social conscience of all South Africans by assisting to sway popular opinion onto ecological and humanitarian grounds.

J.W. Salem (Acting Chairman and Founder)

"What is this!" Buck asked.

"How do you mean?"

"I thought you were going to start an environmental group."

"This is an environmental group."

"What's this then: social problems in South Africa? Social injustice? International law?"

"Well, since many environmental issues here are related to the way people live, I thought FACT should include social issues."

"You mean," said Buck in a disbelieving tone while shaking his head, "expose apartheid problems? I'm referring to this phrase here," he pointed at the page. "Abolition of apartheid. Assisting to sway popular opinion..."

"It's really not a political organization. We will deal with factual issues related to the environment."

"Not political..." Buck said quietly, whistling and chuckling to himself. "Don't you think you're being a bit naive."

"But I want to do something."

"That's commendable, *muishond-kind*. But go into it with your eyes open. This is not a free, democratic society, you know. You or I can't simply start up anything we want. There are major restrictions. Censorship. Many things in this country are set in concrete."

"Look Buck, I know everyone is talking about this nationwide State of Emergency. But things in Hermanus haven't changed. Granted on the way here from Cape Town in the bus I passed Crossroads and that township looks like a burning war zone. Army all over the place. It's frightening. But this is Hermanus and FACT is an environmental organization."

"Don't kid yourself that things aren't different here in Hermanus."

"What do you mean?"

"No, I tell you what. Firstly, you tell me exactly how FACT is going to help?" Buck felt at ease with the formality of this document, holding the page lightly in his hand. Amongst his people Buck was considered a *kwinkslag*, a wise-crack, and where politics was concerned he was a sharp self-educated reader in the classics.

"FACT may not help immediately. But it'll be a start when grievances, whether environmental or social, need to be collected and published. At university I spoke to academics who know of other liberal organizations that may take this further. And since Hermanus *Coloured* Local Authority, since most *non-white* local authorities don't seem to care how you live or how the environment is being affected, that much is obvious, I would say you need backing from these other organizations too."

Buck stroked his beard sceptically with forefinger and thumb.

"To start off," Joseph continued, "FACT would need an affidavit from you."

The old man's hand dropped from his beard.

"But I can't make an official complaint. I can't appear in court! Who would believe me: honestly, which *whities* would believe the ramblings of an old *coloured* fisherman? Pah! And you say your organisation isn't political. In South Africa everything is political - everything - down to your stinking shoelaces." Buck glanced at his own bare feet.

Joseph was now sitting stationary in the hammock.

"And this here statement you write", continued Buck: "'to sway popular opinion onto humanitarian ground'", he pointed repeatedly at the piece of paper, getting flustered with the futility of this age-long apartheid topic. "That, you say, is not political? Not political - Pah!"

"Calm down Buck," Joseph intervened. "This group FACT that we're starting is an information organisation, purely and simply that. But it will have a strong conscience. We will visit the *coloured* areas and survey your social conditions in relation to job opportunities, education, alcoholism, crime, malnutrition, slum diseases, and how you rely on the natural resources around here. These facts will then be published for proper political organisations to act on. It's all I can help with at this stage."

The old man shrugged sceptically. "You're playing with fire," he said, shaking his head cynically. Joseph rose from the hammock and playfully punched Buck on his shoulder. Smiling at Buck's seriousness, he flopped back into the hammock again, lying flat this time and facing the mottled bamboo ceiling.

"Okay rock-hopper." Buck cheered up seeing Joseph become more animated. "Now tell me why you came in here looking so dejected?"

But this question instantly prompted another silence. A silence that was punctuated by the squeaking movement of the hammock swinging slowly back and forth, back and forth.

"Is it your father again?" Buck probed on the off chance. It would not be the first time they had discussed the deep discontent felt by Joseph which often stemmed from his father's insatiable demands. He knew of Joseph's sensitivities, his susceptibilities to his father, which most often caused Joseph to withdraw into his own shell and eat himself up with conflict and family guilt. It was a habit, Buck knew, that each Salem family member shared in their own way: Jessica withdrawing into her music; Mrs Salem into her persistent fretting; Mr Salem into his idealism and brute determination to see things one way only; and Joseph immersing himself single-mindedly into his studies. Seldom did the Salem family openly share their feelings. It was like a microcosm of the closed society in which most South Africans lived.

"I can't cope any longer at home," Joseph confessed, looking up at the bamboo ceiling and rocking slowly. "Many things are involved here, not just Dad, Buck. How long must one wait to be listened too?" Joseph turned away towards the wall and the hammock squeaked with this motion. Randy tilted his head and let off a soft dog whine.

"What things? Tell me, *seuntjie*."

Joseph turned back. "My birthday happening tomorrow night. The farm. My science studies at university... but it's useless even talking." Joseph sat up. He raised himself on shaky feet and began to make his way to the door.

"Hey, *wag 'n bietjie*!" Buck pleaded. "Don't go yet. I want to play you something."

"What?" Joseph was holding onto the doorframe.

"Come inside man. Here's a story that might help. A story about my life. Of how some people give more than they get and then are taken for a ride." Buck's face was animated, alive with expression, a sea of moving lines and wrinkles and frowns. "Until one day they make friends with a dog, or the ocean, or this here harmonica."

Joseph walked back to the hammock and flopped limply into it. Already Buck had armed himself with his rusty harmonica, it almost disappeared into the fuzzy grey beard as he put it to his mouth and started playing:

"I..(hum) was bo..orn (humm) under a

wand..ering star (hm hm hiimmm), I..
was bo..rn under a wand..ering star,
hm hm hi..imm, hm hm hi..imm, hm hm
ho..omm, hm hm ha..aomm, I was born"

Immediately the atmosphere lightened. Randy rested his head on two paws and closed his eyes peacefully, while Joseph joined in humming the tune to himself. The old man's cheeks bulged red at each breath as he blew into the harmonica, rapidly and rhythmically moving his shoulders side to side with the beat.

"Okay, alright, that's enough now, Buck," Joseph started to become impatient when the tune began to drone on. "Now what is your story?"

"That is the story!" Buck glowed with achievement.

"What?" Joseph's tone was incredulous. "Don't play games with me."

"I tell you *muishond-kind*, it's those words of the song." Buck lowered the harmonica with a serious expression on his face. "Do you remember when we watched that film 'Paint your Wagon' in the hotel?"

Joseph nodded.

"Remember Lee Marvin as the wandering gold miner?"

"It's a little vague, but yes."

Buck grinned mischievously: "Well he's just like me. Like a wandering fisherman. Or like an albatross which flies the Southern Ocean."

Joseph looked at him confused. "Stop playing games, Buck. I still don't know what you mean and anyway, what's it got to do with me?"

"You can't see. Can't you see what I mean?" Buck pushed his chest out flamboyantly as a joke and added under his breath: "With all your years at university and you still can't see... HAW, HAW!" But he sensed the degree of desperation in Joseph's eyes. Supplicatory, pleading eyes. Buck became serious. "My Josef, what I'm saying is that I'm my own person, nobody else's. That is what I mean to be born under a wandering star. It means that I can wander as I please, no matter who demands this or that. Sometimes in life you have to discard people or passbooks along the way... so you take with you a dog or harmonica for a bit of company when it gets lonesome."

Joseph sat absorbing these images. Of a wandering fisherman on the beach. Of a wandering albatross: free, soaring with majestic wingspan, like the stuffed museum specimen he had hung in a steep dive near his bed. Then he imagined himself as a wandering scientist, and the vision seemed clear. Somehow in this fisherman's cottage the vision of the scientist was acceptable.

"You make it sound too simple," Joseph declared.

Buck held up his knotted hands in exclamation: "But of course it's simple. Look around my house here, how simple life can be!"

Joseph scanned the sparse interior.

"Come with me..." Buck stood up and led the way. "Now you said that the State of Emergency has not yet arrived in Hermanus. Come, I want to introduce you to someone."

*

They crossed a fence bordering the *Coloured* Quarter. Beyond it was an open space of *veldt* of knee-high grass in which grasshoppers sprang to life, whirred up and away in flight, as Buck and Joseph walked through. The *Black* Quarter towards which they headed was economically of a lower class than the *Coloured* Quarter from whence they had come. Apartheid, with its racial spotlight, legalized even the separation of the *coloured* race group from *blacks*. As Buck led the way with an easy stride, humming to himself, Joseph contemplated the wisdom and experience of Buck's simple instruction. They approached a shanty township of one-roomed dwellings constructed of corrugated tin and wood. Buck located a breach in the surrounding wire fence and they both entered a shady narrow corridor separating one row of shanties from another.

At what resembled a condemned shanty, a small shack of loosely nailed panels of corrugated tin, its wooden frame moldy and rotten in places, Buck stopped. The shack lacked windows. They circled it to access the tin door which revealed an unbolted padlock on the outside. Rocks had been finely balanced on the perimeter of the roof, weighing the paneled structure down. Buck knocked on the shaky door. A man maneuvered the entrance open and they were admitted. It was dark inside without any form of natural light, the only available light emitted from two candles. Buck introduced the man in the shack as Simon Thandiwe.

"Hamba kahle brother!" Simon smiled and gave Joseph a firm three-clasp handshake.

"Simon," Buck explained, "has just fled the township of Crossroads. He has come to live in Hermanus for a brief time while things cool down. He may be able to offer you advice."

Joseph felt puzzled. Caught off guard. Suddenly finding himself face to face with a resident who had lived on the other side of the razor wire at Crossroads, the township he had previously viewed speeding past in the bus. Joseph's eyesight still struggled to adjust to the dim interior of the shack from the bright light outside. And the air in this confined space was heavy and stale with a faint odour of fungal mold. Of Simon's dark features, the candlelight glinted off a scar from a burn that ran all the way from his left ear down across his shoulder and neck.

"How can I help you, Joseph?" Simon's tone was warm, his smile was generous, his voice matter of fact.

Buck intervened with an introduction: "This is my adopted son so-to-speak, my *muishond-kind*," he said putting an arm around Joseph's shoulder. "He is a great nature lover. It's in his blood Simon, I have seen it grow from his earliest days. I was first convinced of it when he was eleven years old and Joseph persuaded the town council to issue a legal notice to protect the local flowers and birds around Hermanus. Joseph here wants to become a biologist."

Simon nodded with interest. "So what's the problem?" he asked.

"Tell him about Crossroads," said Buck. "Tell him about the need to move on if necessary when the time challenges. Tell him about following your dreams."

"A-hh, that I can do," Simon smiled and turned his compassionate eyes onto Joseph. "But I'm not sure where it fits in with his biology?" Simon shrugged. "If you don't mind me continuing with the wallpapering over there while I talk."

Buck gestured that he go ahead with the gluing of old newspapers to the walls.

"Why do you do that, Simon?" asked Joseph.

"What. This wall papering?"

Joseph nodded.

"It's for insulation. The walls of this shack are just made of thin corrugated tin which loses heat easily and gets very cold at night. My only heat in the evening comes from that paraffin burner over there. Pasting layers of newspaper, like this, onto the walls and ceiling helps sustain the heat inside on cold nights."

While Simon collected a pile of recent newspapers and spread them on the earth floor, Joseph walked around the cramped interior in the dim light - his eyes now adjusted to be able to read bits of news off the wall.

"How many layers of paper do you put on?"

Simon laughed. "Twelve layers in most places. Onto older layers stuck down previously."

"It's like an archaeologist's dig. I bet if you peeled off this paper layer by layer, you'd probably get a history of South African news going back years!"

Kneeling on the floor, Simon looked up amused as he stirred a bucket of homemade glue. He was beginning the process of applying glue onto the paper with a rag attached to a wooden stick.

"The news as certain papers report it!" Simon corrected.

"True. May I borrow that candle to read?"

"Sure, just don't set the whole place alight."

Joseph approached a section of wall near Simon's single bed raised on bricks, and in the flickering light of the candle held in his outstretched hand to the print, he could make out:

GO AND SEE MANDELA, SUZMAN TELLS GOVERNMENT

The government was today urged to adopt the same stance in favour of negotiation as jailed African National Congress leader Mr. Nelson Mandela.

Two Progressive Federal Party MP's who saw him yesterday, Mrs. Helen Suzman and Mr. Tian van der Merwe, warned that he could be one of the last of the traditional black African nationalists willing to negotiate. They spoke against the background of statements by President P.W. Botha that a nationalist, non-communist section of the ANC could talk to the government.

The Argus, Tuesday 6 May 1986

PEOPLE TREATED LIKE CANNON FODDER - CHURCH

The lives of human beings have become "just so much cannon fodder" according to a leading article in a church newspaper.

Dimension, the mouthpiece of the Methodist Church of Southern Africa, discusses the split in the church over political involvement and comments on escalating violence. The newspaper said that "the level of the violence appears to have intensified as opinions have hardened".

"The lives of human beings have become just so much cannon fodder - with the former chief of police in Soweto, 'Rooirus' Swanepoel, suggesting that more violence should be used in unrest

115

situations and the 'comrades' and other dispensers of mob 'justice' in the townships resorting increasingly to barbaric assassination of anyone unfortunate to cross their paths."

"Its an endemic cycle of violence that bodes no good for the future - short or long term."

"It represents a stagnation of human values that cheapens life and threatens the very existence of all men and women."

"It is sinful."

The church responded with dismay to the cycle of violence. It condemned violence no matter who the perpetrator.

The Argus, Friday 9 May 1986

"My dream is simple," Simon began. "I just want to see my people free."

Joseph approached Simon to share the light from the candle he held, as Simon pasted wet newspaper sheets to a wall, then smoothed them down, three sheets thick at a time.

"And yes," he continued, "you Joseph, and Buck, are amongst those people. Everyone in this apartheid country needs setting free. I'm what they call a political activist. Thandiwe, my surname, means 'beloved'. I have set myself the task of being a political educator wherever I go: to teach people to realize and to demand what is rightfully theirs. For this my people feed and clothe me and give me a shack, as they have done here, or take me in while I move around the country conducting these illegal underground activities."

"So you know what happened at Crossroads?" Joseph asked.

He shrugged. "The shanty city of Old Crossroads just mirrors the tragedy of apartheid poverty. A questionable black leader became corrupt through backing by the apartheid government, became power-hungry and dictatorial. And when people turned against him, he brought his *witdoeke* vigilantes onto the street to maintain his power. Power that was far removed from the people. It led the vigilantes, many say assisted by police and the defence force, to eventually burn the places of opposition to the ground. In fact, Mr. Le Grange, Minister of Law and Order, refused to oppose a final order restraining the State and the vigilantes from attacking. And now the Crossroads Executive Committee have sued Le Grange for damages: although how does money compensate for the destruction of an entire community?"

Joseph listened intently.

"See there," Simon pointed. "Read those newspaper articles I've purposefully pasted on the wall next to my cot. Each day when I wake they will inspire me for my work here in Hermanus."

CROSSROADS: WHO? WHY?

The apparent impotence or, according to reliable eye witnesses, resolute unwillingness of the authorities to protect hapless refugees and other law-abiding citizens from marauding gangs of vigilantes in KTC, Crossroads and environs is assuming the proportions of a national if not international scandal... The burning questions remain. Have some 50 000 people become homeless as a result of a deliberate decision to exploit communal tensions and promote functional hostilities? By whom would such a decision have been taken and executed? To what purpose? For the sake of counter-insurgency operations against the so-called 'comrades'? Or a forced mass removal to Khayelitsha? Is this a new and particularly ghastly technique of removal? Neither police nor military have been effectively deployed to keep the peace. Why has this been so? Has someone made a ghastly error of judge-ment? Was this a deliberate, carefully considered policy, supported at a high level? The public is entitled to know...

Cape Times, editorial

ORDERED OUT

Police have forced about 120 Crossroads refugees to leave the Temple Israel synagogue in Wynberg, Mr. Robin Carlisle of the Progressive Federal Party told the President's Council.

The synagogue was raided on Wednesday night and the Jewish Relief Committee was ordered to remove the refugees - all women and children - by June 23 because they were infringing the Black Urban Consolidation Act, he said during a debate on the Internal Security

"A-hh, now I understand you, Buck." Simon chortled.

Buck nodded and winked back at him.

"There are some similarities." Simon was amused. "I am used to dealing with people who have next to nothing: *black* people, not *white*. But Joseph, if you are having to give up everything that is important to you - your biology, your talents, your soul - for someone else, then my advice is to fight for it. There are so many restrictions placed on people in this beautiful country in which we live: we just cannot afford any longer to allow restrictions by other people to destroy our dreams."

Joseph felt a tingle of gooseflesh touch his skin at the simple power of these words.

"I don't know a lot about biology," Simon continued. "Except that in our African tradition animals can teach us a great deal about ourselves. I hope that is the way you will choose to study animals, Joseph: to learn more about us, and to gain a deeper respect for both sides.

*

Leaving Simon in his shanty township behind, Joseph walked as one with Buck, shoulder to shoulder, re-entering the gravel road of Hermanus *Coloured* Quarter with its brightly painted two-roomed houses. The Labrador who had dutifully travelled beside them without ushering a sound suddenly barked and sped off after a mongrel dog up the road, and the fisherman's shouted attempts to discipline and retrieve it were to no avail. They watched it veer away.

"How did you guess I was having trouble with my father?" Joseph asked. "Since I never revealed to you what was bothering me."

Buck smiled: "You didn't have to reveal it."

"I'll see that tune *'Born under a wandering star'* in a new light now," said Joseph. "Thanks for the advice and for introducing Simon Thandiwe."

The old man put an arm around Joseph.

"What are friends for?"

"You won't forget my birthday party tomorrow night, and to remind the others to come along."

Buck puckered his lips: "Wouldn't miss it for the world!"

"Oh, I nearly forgot. Dad says you still owe him some money, but isn't the payment date next Wednesday?"

"Ja... ja, he mustn't worry. He'll get his money as always."

It was mid-afternoon by the time Joseph puddle-hopped his way back out of this squalor under a sky flecked with white wisps of cloud. In an introspective mood he began his route in the sun back to *Rus en Vrede.* This time his thoughts distracted him from the scenery.

Imagining, in long vivid images, the practicalities of Buck's and Simon's words. Wise words indeed. He imagined an albatross, wings stretched to the limit, wandering the Southern Ocean, as he had seen albatrosses do many times. He thought about a fisherman like Buck wandering around the trawlers in Walvis Bay. And he imagined himself as the scientist conducting experiments along the sand dunes of the Namib Desert. These visions were so simple. Terrifyingly simple. Because for all of them: the albatross, the fisherman, the activist, the scientist :   all protagonists were alone.

Were they happy? Yes, he envisioned them happy. Doing precisely what they wanted. Without anchor, without worry, beaming a broad smile like the one reflected on Simon's compassionate face. As well as seeing the world with Buck's humour. They were able to take-off to any destination at their every whim. And then Joseph recalled Buck sitting like merlin in his cave: a space as sparse as Buck's little house, with a hammock as a bed, one lounger chair, and his fishing nets drooping from the ceiling. Is that what life is all about?

Could you really substitute a harmonica for, let's say, another human being? Buck did once mention he had had a wife but Joseph doubted him ever describing children or a family. Where would Lexa fit in with this wandering albatross image? Indeed, where would his father, mother, where would Jessica fit in with it? Somehow Joseph had never envisaged his career being such a lonely path.

Did he have to choose simplicity like they had done to achieve what he wanted? And what exactly did he want? Joseph's thoughts were flowing as rapidly as each footstep.

- CHAPTER 5 -

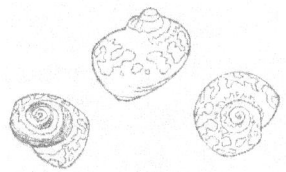

It was an odd feeling, a sense of unreality, of separate world's split apart, fragmented and closed off, when Joseph lifted first his left leg, then his right, high enough to step through the dense *acacia* hedgerow. This barrier of bushes isolated and effectively blocked off the *Coloured* Quarter from view of the affluent white suburbs. Less than a hundred paces on the other side of squalor which Joseph had passed quickly and quietly out of, were plush and luxurious homes, many double-storeyed, nestling on trimmed lawns being irrigated with sprinklers that pulsed arcs of water in slow turning circles. As he arrived back at *Rus en Vrede,* mid-afternoon tea was being served on the porch.

"Is that you, Joseph?" Mrs Salem called out when he closed the front door with a frustrated slam.

"Yes Mother. I'm going up to my room." This answer, lacking his usual patient affability, she found puzzling.

"Won't you come here for a minute, dear." She was pouring tea from a saffron coloured china teapot when Joseph stepped onto the sunlit porch. Mr Salem received a refilled teacup as she brushed away biscuit crumbs from his blanketed lap. The man gestured a greeting from his wheelchair and Joseph couldn't help but warm to the sight of his father's unkempt, rather bushy-haired appearance.

"I reminded Buck about the money he owes you," Joseph said. "He promised that you'll definitely get paid next Wednesday." Mr Salem looked up, acknowledged this with a smile, then dunked a brown ginger biscuit into his teacup.

"So you visited Buck?" Mrs Salem enquired.

"I saw him, yes." Joseph responded curtly, a reply that helped him to veer away from raising suspicions he had visited the *Coloured* Quarter. His father's earlier warning to avoid the short cut to Fernkloof Nature Reserve by transiting the *Coloured* Quarter was now a serious obstacle to his everyday movements. Notwithstanding the State of Emergency security implications involved, today he had ignored this parental restraint, perhaps just one more time, and managed to elude the beady eyes of Wouter Thyssen out of sheer necessity to talk with Buck. In future he would probably have to avoid the *Coloured* Quarter.

"You missed our delicious buffet lunch: cold meats, potato salad, French salad, fresh fruit and ice-cream. You didn't forget lunch did you?" she asked with a degree of concern. Joseph realized that indeed he had forgotten to eat; those stomach cramps had burned away any desire for food. "But you're looking much chirpier than earlier on!" she said with relief in her voice. "Must have been the fresh sea air that did it. I'm pleased you're looking better. Grab a bite in the kitchen if you want, and we'll see you later then."

Ascending the stairs Joseph wondered whether either of his parents had ever really experienced at first hand the true squalor of the Hermanus *Coloured* Quarter. Surely they had noticed it driving by. But had they ever dared to understand it? Understand what was perched on their back doorstep? Often enough Joseph had noticed the extreme poverty, but it never affected him as much before as he now felt disturbed by it. As if this nationwide State of Emergency was forcing him to examine apartheid in a new light. In a more personal light. And apartheid, beyond the squalor, seemed even more to be a poverty of the soul. Forcing people apart. An era of unfulfilled talents. A political reign of lost dreams. Normally restrained and taciturn and disinterested in politics to any great extent, as were most scientists, Joseph felt surprised at the strength of his new feelings.

In silent contemplation he stopped three quarters way up the stairs. Thinking that apart from the extraordinary ability the so-called *non-white* people had to survive these hardships yet still smile, they were beginning to fight back for their dreams. For some semblance of a future. In a way – in this conflict with his own career - Joseph felt he shared a similar dilemma. A dilemma of broken dreams. Buck and Simon had instantly noticed this similarity by offering consolation and their advice and encouragement. That he may have to fly in the face of his father if needs be.

"What about the starving masses in the rest of Africa?" Mr Salem had often replied, when Joseph pointed out *black* and *coloured* poverty as they drove past a shanty township. "In our country," his father had reiterated, "the *blacks* should be grateful to have any roof over their heads."

As if a shanty roof resolved in his father's mind that this was enough, a propitious enough solution, one that he would certainly not lose sleep over. For the first time Joseph was realizing how different his ideals were to his father's. And how ensnared Joseph felt in a parallel predicament. Feeling as if he too were less than fortunate, certainly less than eager, to be someone in this vast and beautiful country who aspired to have a dream.

*

A Namaqua chameleon had escaped from the thickly vegetated glass tank beneath his bedroom window and was climbing up a curtain. Joseph gently unclasped its claws from the calico material, stroking the agile reptile on the abdomen as it fixed its eyes on him, and placed it back next to two others. Transported home yesterday evening, these desert chameleons were collected during his study at the Namib research station and would later be humanely studied by other researchers. These cute animals would spur him on, Joseph had decided. They would inspire him to complete his chameleon article this weekend before returning them back to the university aviary.

His walk with Buck in the *veldt* between the *Coloured* and *Black* Quarters had served a secondary purpose too. Joseph now unstrapped his hip bag and extracted first the bottled scorpion pickled in vinegar, and then a plastic container. The content of the container was noisy and alive. Filled as it was with green and brown hopping insects: grasshoppers caught by hand returning from Simon's house with Buck. To capture them live, Joseph had stalked, like in his fantasy of that prodigal baboon in the leopard-encounter poster, from an angle where the grass-blade on which the insect is perched obstructs its view of an oncoming predator. Then with a swift downward sweep of a cupped hand the grasshopper is snared, often as it rises in clicking flight.

Removing two insects from the container, Joseph delicately held one in each hand. He was careful to grasp their hindlegs between his forefinger and thumb. Ironic, really, he felt, to introduce them so gently to their fate. But nature, he knew, is unsentimental. Wiggling them about at the top of the open glass tank, each of the chameleons were enticed to eat. The first chameleon began to rotate its eyes, focusing one eye, then the other, together in binocular vision to judge distance. The next chameleon did the same. One by one they aimed a long purple-gray tongue. Flick... flick... in a blur. It did not take long before the five chameleons had consumed a dozen live grasshoppers. Joseph turned his attention to the injured sunbird in the cage by the rear window, then finally he scattered floating food for the Malawi cichlids in the large fish tank.

Settling down at his desk, he opened the 'Chameleon Data File'. His research for publication would contain an elaborate set of five graphs and three tables of statistical figures. Professor Harris had hired a scientific illustrator to draw a sequence of artworks of a Namaqua chameleon in the process of changing its body colour, and afterwards Joseph had added zones of temperature. The illustrations were complete. The article now needed a conclusion and a final summary. Joseph skimmed through the text, sat back, sucking on a pencil.

Usually ideas came to him naturally with little effort. Sometimes new theories, new directions of thought, new conclusions, that often swamped him in a flash. This time, however, a stomach cramp flooded in as three spasms of excruciating pain in close and rapid succession. Joseph gripped one arm of the chair. The final cramp was severe enough to fill his eyes with tears, he gasped at the stomach ache, his head falling forward onto the open file in a grimace. After sucking in air Joseph's reaction was to hold breath but it helped more to breathe deeply, letting breath out then in, out then in, with a controlled, sustained effort. The last few days experience had taught him that much.

Joseph checked his wristwatch. Damn, he silently cursed! It was just a quarter to five. It was too early yet to take more anti-spasmodic pills. Standing up still bent over forward in half a jack-knife he slowly crossed to his bed from the writing desk. Ever gently, grimacing in light fast breaths, Joseph lay himself down flat and then all went dark.

Twenty minutes passed before he awoke with a jolt. The cramps had transformed into a diffuse but bearable backache. Joseph prodded his abdomen, it felt tender, but he was able to sit up. What happened, he pondered? Propped up against the continental pillows of the bed, staring ahead blankly, he was confused. He felt some hunger pangs now ontop of a degree of nausea and fatigue. Lexa, he knew, would be arriving soon. I've got to resolve this now, I've got to, he urged himself!

Joseph tried to put his mind feverishly to work for an answer but only thoughts of Buck re-emerged, recalling what the old fisherman had said. A light euphoria again descended just imagining that scientist wandering alone over sand dunes, free as the flight of a soaring, wandering albatross. Buck's simplicity was contagious: the vision itself brought with it a measure of relief over and above his nagging uncertainty.

But would I be living the life of a hermit? This notion caught Joseph unawares. Imagining a lonely hermit! He balked at this sharp criticism of Buck. Am I starting to criticize everyone, he wondered? Next I'll alienate myself completely! He clenched his fists which helped quell the abdominal discomfort and the unease in his muscles which were tense in his lower back. The pain and confusion resembled what had happened at university residence, his peace of mind clouding over with emotions again and he felt a dire need for the solace of a friend like Buck. It was just so easy to fantasize that his family would understand.

Talking to Dad now seems out of the question. It was all he could do to admit this to himself inwardly. Mother is bound to align herself to one point of view: to Dad's side, the opposite side, and by doing this, in her conflict, she will become emotional. And neither does poor Jessica deserve this conflict, she usually ends up sinking further into her world of music. I'm becoming an island unto myself, he admitted! How different is this to the hermit? An early sign of another stomach gripe, a slow one, began to make its presence felt. Joseph remained calm by clearing his mind for a few minutes trying to think of nothing: rhythmically and forcefully inflating his lungs, then relaxing.

He needed to write Buck a letter. Increasingly that image of the wandering scientist touched him with intrigue. So too that tune '*I was born under a wandering star.*' Why did Buck describe my future in this way? After all, Buck hardly knows about the doctoral offer at university. Joseph felt he deserved more feedback, a more specific conversation and a friendly receptive ear with whom he could reveal the precise nature of these uncertainties which he'd been unable at first to express to Buck. Or else these cramps would continue turning their vicious spiral.

On a piece of computer paper, he drafted a note and folded the paper into four, on one outer fold capitalizing the name: BUCK WILLIAMS. Joseph glanced up at that poster of the leopard-baboon encounter. The action blazed out in a cloud of dust: a life and death skirmish between two animal adversaries on an unknown salt-pan in South Africa. How had that skirmish played out? Who had survived? Slowly Joseph rose from his bed to begin a walk to the door holding onto his tender abdomen.

*

By the time he reached the entrance to Hermanus *Coloured* Quarter, the shadows were long and drawn, the mud potholes and puddles of sludge on the dust road and silhouettes of refuse were blackened in the light. This road leading up to the poor houses resembled a landscape on the dark side of the moon or the scene of some fierce light-artillery bombardment. Joseph decided to obey his father and not enter the township.

Unfolding the note, he reread it:

Dear Buck,

I need to talk to you properly about how I feel. This afternoon I hardly said anything and yet you presumed what I ought to do: you know, '*Born under a wandering star*' and all that. It sounded feasible to me when you first explained it but now other things are in the way. I even thought what it must be like for you to live alone. It's fine to be able to wander whenever or wherever you want, but why should this freedom mean you must be alone?

Joe

It took several minutes before Joseph could hand over this note to an elderly resident of the *Coloured* Quarter who happened to be passing through the entrance, and who, with a broad toothless smile and a tipping of his hat, said it would be "a plesier meneer" to deliver the note to Buck's house.

The sun cast a shadow behind Fernkloof mountains as Joseph headed for the promontory to watch its golden glowing disc touch the sea. He could walk more briskly

132

now, his abdomen relaxing with the stride. At *Kafferklip* lookout the sun had billowed into a red fireball orb, the kind of deep crimson ball that leaves a glowing imprint on your retina when you quickly glance at it and then look away and blink.

Before sitting down at 'Ruth's bench' Joseph paid his silent respects by reading the plaque secured to the backrest of the bench:

'With love we remember RUTH, whose gentleness,
compassion and courage endeared her
to her many friends
1953-1981'

Joseph sat down and allowed the pleasure of the sunset to wash over the day's problems. The depth of satisfaction he felt, gained from this panorama, was as amazing as he ever remembered it to be. This taste for nature which he had developed was as resolute as it was delicate, a taste refined through a lifetime observing nature's course. Of a life spent outdoors. Observing it. Interacting with it. The depth of pleasure he felt was unlike many human-made satisfactions: cultural pleasures often lacked the depth or proportion of feeling, often lacked the timelessness and expanse of sitting in a place where nature touched so many unfathomable dimensions to the soul.

His stomach cramps were gone. The exertions of the day had left their toll in the form of a dull throbbing headache. But by gazing out at this sunset panorama all conflicts were drained. The release felt no different to when Joseph worked on his journal. With the exception of Lexa he had not really told anyone about this journal exploring his life's work, so there were no family expectations attached to it. No abdominal gripes to contend with, no emotional links or ties. Unlike the university 'Namaqua Chameleon Project' which had earned him the distinction of a doctoral position.

Lexa, he remembered, may already be waiting for him at *Kwaaiwater*. The crimson sun was dipping deep below the ocean horizon with more than half its diameter sunk. Distant on the sea surface were shimmering spangles of light. The tips of three long trails of cloud were burning a bright gold colour, the shoreline was bathed in a tangerine glow. While standing and facing seaward Joseph watched the brilliant semi-circle diminish to a phosphorescent point, then dissolve into nothing, and he decided to take the longer, lonelier path home.

*

Lexa had telephoned. Joseph's mother had jotted down her message informing him Lexa would be half an hour late.

"She's been delayed on the coach home from Stellenbosch university," Mrs Salem explained further. "She needs to shower first and change her clothes before coming over." Selfishly, self-indulgently, Joseph imagined that Lexa may also be writing him a birthday card and wrapping a birthday present. He wondered what it might be.

Upstairs, he avoided the 'Chameleon Data File' and took out his journal with the reinforced red cover from its safe place locked in a drawer of the desk. The journal bulged with A4-pages. Originally he'd planned it to be a skeleton draft of his life's work: '*The Evolution and Basis of Human Consciousness*'. All extraneous theorizing and suppositions and myths would be kept to a bare minimum. Only relevant scientific facts would be listed, in point form and in pencil. But somehow Joseph's enthusiasm for the subject had transformed the skeleton draft into a tome of 342 pages. It needed to be edited now, re-worked, in many places entirely rewritten, because ever since proceeding he had matured in the process of writing it. In fact, he doubted whether anyone else would be able to interpret these notes. It made him eager to begin a rewrite: perhaps there would be time during the summer holidays.

Joseph paged through the journal balanced on his lap on the bed. The latest entry was dated a fortnight ago. He sighed, closing his eyes to the pleasure of resuming this work, this journal, old soul mate to his life's work. It was a subject which struck him so early in life. As if he'd always known it inside of him. As if it was always part of his life. He approved of the word: 'life's work'. That made sense to him. The pleasure gained doing it wasn't the kind of pleasure experienced when struggling to solve a work problem and eventually succeeding. Because that was more like relief after a forced effort. No. To Joseph this challenge was all absorbing. Truly a life's work. Commenced at an early age. Gripping his concentration until everything else blotted out. Once he had described this drive in him, this appetite, this restlessness of his talents, in terms of love.

"C'mon Joe," some would say, "you can't work all day and all night!" His mother questioned him early on in his university days: "Why don't you meet your friends more often? Take Lexa to a movie? What about joining the Hermanus sports club?"

Eventually they tired of asking.

"Mother, the way I see work isn't how other people seem to see work. It's nothing to shy away from, it's not something reserved for particular hours of the day during which time the person involved exists trying to kill time, or struggles through wanting to go home, or plans the next holiday in their head. Work is my play. So why should I limit the time I spend on it, or whinge and complain about it, defer it or search around for the pleasures other people call play? Their play is their play. Only sometimes is their play my kind of play!"

People finally conceded this explanation. They were sceptical at first, disbelieving of the almost inhuman hours of work Joseph involved himself in. During the years at *Kwaaiwater*, as Joseph's dedication deepened, Mrs Salem decided not to let on to her husband the course these developments were taking. It was unnecessary to tell him anyway. Mr Salem's immobility prevented him from ever going upstairs, or following Joseph outside, or he would have realized how much Joseph's interests had burgeoned.

In his journal, at chapter 18, Joseph had reached the subject of human instincts and drives. This would be a chapter comparing human behaviour with new findings on the behaviour of other mammals. In the way he was tackling it, it was valid to compare similar behaviours of humans with other mammal relatives. By doing this Joseph was assuming the role of an ethologist: a scientist in search of the evolutionary links in animal behaviour. By comparing human and animal behaviours, he may be able to unravel and understand why and how certain types of human drives take place.

At this point in the study Joseph was monitoring how instinct is dampened, controlled, how it is subdued when an animal is domesticated. This would help him establish any links as to why human instincts under the domestication of living in crowded societies may also be dampened. The animal he was busy studying for comparison was his Siamese cat Mowgli. Siamese cats were an ideal breed for this kind of work: they are among the most intelligent cats, sensitive, highly strung, and they can be remarkably emotional. In the past four months Joseph had managed, unusually for a cat, to train Mowgli to react to a few spoken commands.

Tonight he would begin to test the cat's instinctive response to the new Namaqua chameleons, establishing the cat's control over these drives, and then the experiment would be repeated with ordinary laboratory mice. Joseph clipped coding sheets onto a clipboard on which he could jot down the cat's behaviours in a quick coded form. He called Mowgli off the windowsill. The cat arched its back in a languorous stretch and jumped down onto the polished oak floor.

"Stay!" Joseph commanded in an even tone. Mowgli stopped, its ears pricked forwards, then sat motionless in the middle of the room. The cat's speed of reaction to his command was timed with a stopwatch, its type of stationary behaviour written down in coded form. Moving to the large glass tank filled with plants and thick foliage,

Joseph lifted out one of the Namaqua chameleons. He placed the reptile onto the floor a distance away from Mowgli. The chameleon began a jerking forward locomotion heading towards the bedside table. Immediately Mowgli's ears swiveled, nose sniffing silently at the air, but the cat's attack instinct was slow to arouse. Mowgli merely moved his head up and down trying to interpret the unfamiliar movements of the strange reptile. Joseph picked up the chameleon, handled it delicately to make it more active, then placed it back on the floor, this time stroking its hindquarters. The reptile broke into a run.

Mowgli stepped forward.

"Stay!"

The cat half-stepped forward again but dampened its response due to Joseph's voice, and sat still.

Reaction time was noted.

It was the first time Mowgli had ever hunted a chameleon and yet his drive to attack was easily muffled. The ability to subdue the cat's instinctual drive through a trained command showed that Mowgli had a degree of control, perhaps a level of consciousness, over its attack actions. That was the real aim of this set of experiments.

Joseph picked up the cat to put it once more at the centre of the floor. With due care he placed the chameleon back onto a branch in its glass tank, before selecting a black and white mouse from cages beneath the dissecting bench. The tame mouse was positioned at the exact spot where the chameleon was set free. Immediately, the mouse darted forward and stopped.

Mowgli stepped forward.

"Stay!"

The cat took another step.

"Stay!"

Mowgli sat.

The mouse ran a distance and froze.

Mowgli bounded two steps, his eyes and ears alert, his spinal hair raised in excitement and arousal, his tail bushed out and waving from side to side.

"Stay!"

The cat's feline concentration was fixed on the unpredictable mouse.

"Lie!" Joseph commanded.

Mowgli crouched low at these words, but then inched slowly forward on his belly. There was now considerable ambivalence in his movements, his instinctual cat drives were competing wildly with his training: Mowgli inched forward, then eyed Joseph, then inched forward again. The threshold was close and soon to be reached.

Promptly Joseph picked up the mouse, returned it to a cage under the dissecting bench and moved over to stroke and comfort Mowgli. The cat was utterly on edge: volatile, wide-eyed, yet it had managed successfully to repress its own instinctual drive to attack. He knew that Mowgli must rest: there could be no more behavioural trials for a while now. Anyway, Joseph knew he had gathered sufficient data for this first test. Beneath the coded behaviours he had gathered on the cat which he'd rapidly jotted down with reaction times in tenths of a second, Joseph wrote:

EXPERIMENT 86 (TEST 1) NOTES

Laboratory mouse chosen of same body-size as Namaqua chameleon. I have tested Mowgli's reaction times to two modes of chameleon locomotion (slow walk and sinusoidal run). The chameleon run elicits more drive-to-attack than the chameleon walk, with the attack threshold markedly higher than for the mouse. This test still to be repeated on three other chameleons at least four times. The cat may be reacting to chameleon movement, body shape, skin colour and/or skin odour.

1) Movement and body shape: Construct a range of chameleon-shaped models and mouse-shaped models. Simulate chameleon movements in the mouse models, and vice versa, including intermediate shapes.

2) Skin colour: Washable watercolour paint to be used on live chameleon skin to mimic mouse colours (black and white blotches).

3) Skin odour: Rub soiled mouse bedding onto the chameleon skin which, by my accounts, is normally odourless, but chameleon skin odour must be gently tested.

4) Animal movement, body size, skin colour and the skin odour variables to be tested with sufficient repeats in all four combinations.

*

"Hello Joe."

Lexa stood at the door leaning against the doorframe. With head tilted mischievously to one side she smiled; a gift held in her hand. On her feet were leather sandals and her chestnut-brown hair was loosely tied back. Around her neck was a sprinkling of colourful beads, the few cyan beads in the necklace highlighting her eyes. She was dressed in an easy-fitting skirt the colour of young green *veldt* after the rains, and a white T-shirt with raw-hide leather belt trimming her waist. She extended the gift towards Joseph.

"Here's your birthday surprise."

He met her halfway across the room and they embraced lightly. Lexa was sensitive not to cling, so she pushed away with a gentle touch and handed over the parcel.

"For me...?" Joseph joked.

The two-toned wrapping paper was as simple and stylish as her outfit and inside was a copy of the great Konrad Lorenz's book: *On Aggression*. On the front page of this influential book about the evolution of animal behaviour, she had inscribed a birthday message.

"You remembered!" Joseph said, kissing her cheek. "I've read it, you know that, but for so long I've wanted a reference copy. Thank you very much."

She nodded with a happy glow at seeing his child-like joy at receiving the book. Lexa commented that she was eager to see the photographs taken during his two week field trip into the Namib desert, but agreed with him that it could wait for later. Mowgli was grooming on the floor, so she picked the cat up and started it purring.

"Were you observing Mowgli again?"

"Only just started. Also I brought home the Namaqua chameleons we collected out in Namibia." Joseph pointed to the large, vegetated tank on the floor, and she walked over to it and knelt down. There was so much foliage and so many branches within and rising out of the glass tank that it took her minutes to locate one of the camouflaged reptiles. They had turned a pale emerald green, the colour of the leaves, and were perfectly stationary in their camouflage.

"They're quite a bit bigger than the dwarf chameleons we get here in Hermanus."

"Yes, a lot bigger. Believe it or not those three over there are still juveniles. I chose them small, about the size of my black and white mice, so that Mowgli's reaction could be compared to both."

Joseph picked up one of the smaller chameleons and its clawed hands gripped onto his fingers. The little hands were powerful, the claws pricked his flesh. The reptile rotated its eyes to gain a view of Lexa as she touched, with a hesitant finger, the dry yet soft scaly texture of its skin. He replaced it in the tank.

"We really ought to go downstairs, Joe. Your mother said dinner was almost ready." Lexa helped him put the clipboard and datasheets neatly next to his computer, then she led the way.

- CHAPTER 6 –

Gritting his teeth, biting back the taste of an impending showdown because it was to be the first extended meeting with his father since breakfast, Joseph followed Lexa downstairs. He felt tense. Lightheaded. He reminded himself that this apprehension was probably worse than actually facing his father in person. Which he must do, he knew. I shall try to be diplomatic, he thought. Firm, yet diplomatic, I shall attempt not to get the man riled. Neither must I spoil the evening for Lexa, he promised himself, as Joseph ran his fingers through his hair thinking that being a scientist, like the art of diplomacy, meant being prepared and learning patience and timing.

To him it was strange how these silent contemplations often preceeded weekend meals at home. *Rus en Vrede*: Rest and Peace! Joseph laughed out loud at the irony in the name, a laugh which caused Lexa to turn around surprised. But Joseph merely winked back at her and steered her onwards. They passed through the entrance hall darkness like two ships sailing from one lighted port to another. The dining room was brightly lit by lamplight. Inside it the large rectangular dining table was place set for five, with a sound emanating from the kitchen: a tinkle of plates being moved about.

"Do you need help, Mrs Salem?" Lexa raised an enquiring voice above this commotion in the kitchen. Before any answer was received, she left Joseph's side to disappear through a doorway. Mr Salem was sitting quietly, a lone bowed figure at the head of the table.

"Joe m'boy, come sit next to me." He gestured by extending his powerful left arm, and Joseph pulled up a chair beside him. The place setting to the right of the wheelchair was reserved for Mrs Salem: from there she could serve her husband whatever he needed. With a stout man's hand on the tablecloth Mr Salem leaned over, his voice reflective and easy, in a confiding tone:

"I've thought about our talk on the porch."

"Oh, yes?" Joseph sat back suspiciously.

"Our discussion about your chameleon project," he said nodding with conviction. "What you've found out about why those chameleons change colour. I found your discovery quite fascinating."

"You're just saying that!"

"Of course not. Joe, you sound surprised?"

"But Dad, I thought you disapproved of its relevance."

"C'mon, you know me better than that." Mr Salem playfully picked up a red serviette and tossed it at his son. Joseph caught the serviette with a smile. It appeared as if the earlier threatening atmosphere established this morning had lifted. Joseph examined the serviette, folded it neatly, feeling relieved. Though there was still a wary sensation in the pit of his stomach: a sort of suppressed, tired elation, pleased not to be discussing his birthday. That would lead to questions about the farm inheritance. It made him only too eager to talk about chameleons instead.

"The really difficult part of the study was how to measure their body temperature. How to get my custom-made equipment to work in the heat of the desert."

"Tell me."

Mr Salem's attentiveness was surprising. Positive.

"Well - now don't laugh OK - I designed and built three little transparent and pliable jackets to fit snug and tight over each of the animal's bodies."

The man's face remained serious; his expression intrigued.

"A chameleon's skin, you know, doesn't necessarily change colour all at once. It changes colour in patches."

Mr Salem nodded.

"We needed a jacket with a designer fit. One that would fit tight enough to enclose both flanks of the chameleon, but also be perforated to let the heat through.

A crucial part in building these little suits was to line them with micro-electrodes; tiny electrodes sensitive to heat. It took me hours and hours just to line each suit with pin-tip-sized electrodes, spacing them apart at close intervals and linked together by a fine meshwork of micro-wires."

Lexa entered the dining room carrying a salad in a wooden bowl. She placed the serving bowl on the table. The maid followed her with a tray bearing two bottles of salad dressing together with a pepper grinder and olive oil decanter. Lexa briefly glanced and smiled at Joseph deep in conversation with his father, then returned the way she had come. Mr Salem sat back, appearing a little weighted with the jargon Joseph was using to describe the scientific equipment, but nonetheless got the gist about the little chameleon suits by nodding enthusiastically.

"We weren't sure," Joseph continued, "whether the suits would work in the field. In our laboratory at university the jackets tested brilliantly, but in the desert the first microcircuit malfunctioned after just one day. The animal wasn't injured, no damage there thank goodness. But the soaring temperatures out in the desert kept melting the electronics..."

Mrs Salem wheeled in the dinner trolley. She had supervised the maid to prepare a fish casserole in a large earthenware dish. Rice, green peas, carrots, a cauliflower cheese, were steaming in separate serving bowls. On the lower tray was a stack of warmed dinner and side plates.

148

Mrs Salem wheeled the food trolley around to the head of the table, positioning it beside Mr Salem's wheelchair, before she sat down. Lexa re-entered the room carrying jugs of both orange and mango juice.

"Jess!" Mrs Salem called out. Emanating from behind a distant bedroom door the muffled playing of a flute stopped.

"Yes Mother?"

"Enough music now. Supper is ready."

Jessica appeared promptly and seated herself opposite Lexa.

"Mmm, smells nice." Jessica peered at the vegetables being dished up by her mother.

"You two were deep in conversation." Mrs Salem eyed her husband, then Joseph, in turn.

"My son here was giving me a lesson on how to measure the skin temperature of a chameleon."

Jessica grinned. "Well now we know who to ask next time in an emergency!"

A general murmur of amusement spread across the table.

"Hold on, hold it a minute..." Mr Salem's tone was serious as he raised a hand to the amused audience. "What I heard sounded fascinating: desert chameleons dressed in little waist-coat gadgets. I presume the reason you made the jackets transparent was so that the sunlight could penetrate?"

"Exactly!" Joseph nodded.

"You know my dear," Mr Salem said turning to his wife, "this son of ours thinks up some weird, pretty weird and strange concoctions indeed, but in their own way they're actually quite interesting." The couple tenderly held hands before saying grace, and Lexa smiled at Joseph before bowing her head, looking radiant. Joseph felt touched by the scene. For him there was nothing quite so warming as these family suppers when the conversation remained light and respectful. For a moment Joseph imagined Buck alone in the sparseness of his fisherman's cottage with fishnets dangling from the roof, but the thought didn't last long. At this moment the wandering albatross image was plainly too depressing to think about now.

*

After supper the family, including Lexa, made their way into the adjoining lounge to listen to Jessica perform a series of piano solos. Ever since Joseph's departure a fortnight ago she had learned a new sonata, and this evening Jessica chose to contrast the classical piece of music with *Honky Tonk Woman* and two jazz-blues compositions. Around her the family relaxed in armchairs: Mr Salem, Joseph and Lexa, each primed with a glass of Cape vintage port in their hands. And at the conclusion of her piano recital Jessica rose to her feet from the piano stool to take a low and rather melodramatic bow to the sound of the family's applause, herself smiling somewhat ironically, triumphantly.

"Now your Namib poem, Joe!" Jessica coaxed him unexpectedly. He shook his head dismissively at her. She was the family performer well acquainted with performing on the piano in front of an audience often considerably larger than this.

"Jess, you're overhyped. You're being ridiculous."

"Come on, please read it!" she urged her brother, pulling him out of the armchair to his feet. As unexpected as her request was, she revealed that she had prepared for this moment by handing Joseph his diary.

Joseph squinted at her; his eyes unappreciative of the predicament she was placing him in.

"I thought I told you this poem was personal?" he said

Jessica pulled out her tongue.

"Alright... If I must," Joseph conceded to the murmur of approval. He cleared his throat. "I don't often compose poetry, but I wrote this on the plane flying into Namibia... which I realize now I should never have shown to Jessica."

He eyed his sister again. She giggled back.

"This poem is dedicated to you, Lexa," he said, looking at Lexa as she sat up surprised. "It's titled: NAMIB SCENE."

> Gripping the upholstery of my seat
> > a surge forward and a tilt

we rise and bank to the left.
The airport control tower suddenly
　diminishes in size to a
　　plaything, tiny and erect;
　very soon tea will be served.

Aloft on the wind –
　far below an eaglet soars
　　and us...
Enveloped in a metal fuselage
　of man's making
　like some swallowed morsel;
We are satisfied to peep down
　on this world through
　　a double layer of glass.

The plane's shadow is dancing
　a wild African jig
　black on the ground;
While from my perch high above
　this cinnamon-sea,
Lines of sand dunes all parallel
　run north with the wind
　　to their death...
　at the river, at the shore.

Once again, with her overdramatic intensity, Jessica jumped to her feet from a kneeling posture and clapped ecstatically as Joseph ended off. While the other three listeners sat in silence absorbing the last lines: 'running sand dunes blowing to their death'. Joseph, for the first time, connected the running sand dunes, as he had written, with Buck's wandering albatross image or the wandering fisherman or scientist. Sand dunes which were blowing to their death. But this image was quickly swept away with the applause of his family and Lexa as they finally reacted, he thought rather over-zealously, to the words of the poem.

"Thanks Joe," Lexa said, still clapping. "I think your poem sings. I think it's very evocative."

Jessica excused herself for the evening, asking to continue with her flute practice. Once she had departed there was a lull as Mr Salem primed his pipe from a pouch of tobacco and Mrs Salem left to oversee the maid clear the dining room before she herself returned.

"Well son..." Mr Salem said adding a final prime of tobacco to his pipe, "tomorrow's your big day. How do you feel?"

Joseph frowned, unsure of the meaning.

"You're nearly legal now." Mr Salem sucked hard as he fired the lighter and an elongated flame shot across the rim of the pipe. "Tomorrow you're your own man. Exciting prospect!"

Joseph hesitated, declining to answer. He was uncertain where the conversation may be heading. Flashing in front of his eyes was a vision of having to sign legal documents binding him to the farm. Lexa, however, remained politely interested in Mr Salem's questioning but equally she appeared oblivious to the real underlying meaning of this casual talk. Since supper she had not spoken a great deal, content instead to listen in on the conversation whilst orienting herself to the Salem family again.

"To be honest Dad, nothing will really change." Joseph shrugged. "I'll be one day older, and granted, also a year older, but that's about it." With this toned down answer he hoped to understate the event and avoid any open confrontation. There dare not be a repeat of this morning's outburst from his father, not now, not if he could prevent it.

"Let me tell you son it's a giant step for us," said Mr Salem, glancing at his wife affectionately, taking hold of her hand. "And you needn't be embarrassed when I say this: that both your mother and myself are proud..."

Joseph lowered his eyes to the floor. Whatever the motives of his father may be: towards the farm, towards Joseph's career and his approaching birthday, the man obviously felt deep affection. He was emotionally moved. He spoke from the heart the truth unto himself.

"Because it's a very special day in our lives, your 21st," he continued. "What a pity we cannot share more of it with you. How unfortunate your room in the university halls of residence must be vacated by tomorrow. But when you return with your bags and things, we will be around."

"Besides, tomorrow night we celebrate your party," Mrs Salem added.

"Surely you must have dreamt about this day, son?" Mr Salem went on. "I know that at your age I did. For me the nostalgia remains strong. Personal memories of the dreams I had, and of family events which I still cherish thinking about. Our family traditions. Our family responsibilities. And that is what I dearly want for you. To enjoy this birthday, Joseph, but also to contemplate and think seriously about your future."

Joseph glanced at Lexa. She listened attentively. Some of Mr Salem's words were wise. But others, Joseph felt convinced, were thoroughly doused in family expectations. Many of his father's words were not original at all - Joseph had heard them umpteen times before – and more than just a few carried ill-considered prejudices about how life in South Africa should be. The monologue touched on numerous subjects. What Joseph could expect of his life, and what life would probably take from him. The importance of respecting his family. The importance of paying one's debts to society and of having to endure a degree of hardship for others. The importance of the *status quo*. The importance of not breaking the perceived rules. It was a mixture of do's and don'ts weaved into a tapestry by a father who obviously cared. Thankfully for Joseph, Mr Salem remained diplomatic enough not to mention the farm in an openly confrontational way. Yet his whole stance, his embroidery of words, created a powerful *motif* around the family inheritance of the farm.

Foremost in Joseph's mind was the issue of self-interest: his father's desires versus Joseph's own needs. It was a prickly subject overspun with what was expected of him as a Salem family member. And with the question which he found difficult to shake off or find an answer to: How can you tell when your parents are right?

*

"Phew, that's over!" Joseph sighed with relief to be leaning over his bed, arranging the large continental cushions against the wall and puffing them up.

"Why are you so dismissive of him?" Lexa asked.

"I'd rather not talk about it."

"But I found your father's words very moving."

Joseph shrugged. "Were they? I don't know if I agree entirely. I just don't want to discuss his words yet. Rather let's hear about your last two weeks." Joseph ducked beneath the model pterodactyl to get seated on the bed. He was holding Buck's note of reply which Franzie, a teenager who lived near Buck, had delivered while Joseph was poeticizing in the lounge and the maid had received it at the back door. Now Joseph stuffed the note unopened into his pocket, in no mood to read it.

"Don't you think you're overreacting?"

"Lexa, please!"

She lowered her voice to almost a whisper: "Do you know why I gave you your birthday present tonight?"

Without waiting for his reply, Lexa took hold of his hand and answered with unusual sentimentality: "So that I'd be the first one to wish you. So that perhaps we can share the passing of midnight together." Her voice quavered with a husky seductive quality. She leaned over, kissed Joseph on the mouth, then slowly drew away.

"Happy 21st for tomorrow."

Noticing a bemused expression on Joseph's face, Lexa sat up straight, waiting for the snide remark. Or a touch of mockery. Joseph was looking upwards at the ceiling.

"Could have sworn that pterodactyl up there woke from its prehistoric sleep and flapped its wings. It must be this sexy birthday heat rising from the bed. You're melting my magnificent predator, Lexa. It's stuck in a dive just inches from bursting into flames!"

"Okay, okay." Lexa wacked a pillow at him. "I get the message." Then smiling playfully, she settled back down. "I'll tell you about me instead." And Joseph was quite happy to listen to her describing her past fortnight, her life on Stellenbosch university campus, the drama exams which she had completed except for the last performance practical in which she would be playing the challenging role of Helena in Shakespeare's *A Midsummer Night's Dream.*

"Joe, will you be able to come and watch me perform? The play is this Wednesday night at *Oude Libertas*."

Teasing her at first with a screwed-up facial expression, then silent contemplation, at last Joseph nodded. "Of course, I wouldn't miss it! And next year is your final year of drama, which will deserve a really major celebration when you graduate."

She squeezed his hand and smiled. "Now you, Joe, I want to hear more about that doctoral post they offered you. And the science medals you've been awarded. When you phoned unexpectedly on Thursday night I thought: at last you've achieved what you deserve. Did it have anything to do with the Namibia trip?"

"Initially I was stunned, you know. Professor Harris summoned me to his office to discuss the chameleon project and without warning he handed over the doctoral envelopes. For his part it must have involved considerable effort to organize the bursary funding and the doctoral post. I didn't know what to say? Yes, working in the desert we did become better acquainted, but whether that had anything to do with it, I'm not sure. He certainly never mentioned it. The difficult chameleon research went very well indeed, so that was probably a main factor."

"Well you deserve it."

Joseph shrugged. "I feel a bit of a fraud being awarded these things. It hardly seems like work, especially

158

according to my father. It's just that it gives me such a buzz of satisfaction."

"C'mon." She nudged him. "Stop being so modest."

"No really, you know how seriously I take biology. How long I've been at it. It's in my blood virtually, even more so since varsity."

She nodded. "So what do your parents think about you doing the doctorate?"

"I haven't told them." Joseph withdrew his hand from hers. He stared absently at the sea windows across the bedroom.

She was surprised at his confession.

"When will you tell them?"

He could sense his stomach beginning to churn with tension, a cramp was beckoning. Behind the blank stare Joseph wondered if he could ever face himself if he didn't find an answer to this question of when and how to tell them. There was no more time available to defer it, or to go outside again for fresh air, or to ask Buck's for further advice. And if he couldn't talk it out with Lexa, then with whom could he?

"Lexa, I feel so impotent," he said. "Powerless to tell them in the light of this damned 21st birthday and the farm." Joseph glanced down, sitting isolated on the bed, his legs pulled up to his chest. Speaking with his chin resting on his knees, he asked: "What's the hell is wrong with me? Lexa, you can't believe what an ordeal I experienced on Thursday and today merely contemplating telling them, and it's not a figment of my imagination. I know how my father will react! But not telling them simply increases the apprehension and makes me feel as if I'm going mad. Ever since Dad's accident, ever since that horrible day, I've dreaded my birthday tomorrow. Between then and now so many insinuations, subtle and not so subtle, have haunted me to drop science for the farm. And Lexa, I realize we've spoken about this before, you and I, many times. But there's no more time left. It must be resolved now... NOW!"

Repeatedly he prodded a finger into the bed.

"Alright Joe, alright," she said with an anxious expression. "Calm down. Just relax."

"A really typical, blatant example of Dad's attitude occurred this morning at breakfast."

"Why, what happened?"

"Oh God, no... my stomach!"

"Joseph are you alright?"

"Wait," he inhaled in deep groaning breaths, as she watched his face with alarm. "I've got to stretch out my legs. I must lie down."

"What's wrong? Are you ill?"

"Lexa... please, quick give me space."

She positioned herself so that Joseph's head was cushioned on her lap and he was laid out flat. Waiting, watching his expression, she squeezed his hand as the fierce grimace of pain on his face subsided. As he lay there gasping softly, she stroked his sweaty forehead; the gentle sound of her reassurances and the feel of her fingers weaving through his curly hair helped to arrest the stomach spasm.

"Phew... that's better. I must relax awhile." Joseph closed his eyes.

"May I talk?" Lexa asked. "You need not answer – just nod yes or no."

He nodded yes.

"Are these stomach cramps serious?"

Joseph shrugged diffidently.

"Have you seen a doctor? Is it a stress condition?"

Twice he nodded yes.

"Joe, you really can't allow these things to affect you like this. Not only is it debilitating, and these cramps mean you're suffering an agony, but this is counterproductive. You're becoming paralyzed by the conflict. Allowing your wild imaginings to take a hold without being in any position to resolve the questions you need answered. Am I right?"

He lay staring up at her.

"You tell me that your parents don't know about this doctoral position. Then how can you judge their reaction? And you imply that you've suffered like this both today and yesterday which gets you no closer to knowing their answer."

"Lexa, I already know their answer. That's why I'm afraid for them and for myself."

She cradled his head, pulling him closer to her.

"Nothing is as bad as it seems," she said soothingly. "In a few months from now you'll probably look back on this day and smile at your folly."

"Perhaps. But first I have to get through tomorrow. You know, Lexa, I was determined not to allow this situation to impinge on the mood of our evening together."

"Then promise me," she insisted, "that at some stage tomorrow you'll tell them! At least let your father know and let him have his say."

Joseph remained in intense thought for half a minute. When this time was up, he nodded.

"Alright. Alright. I'll let him speak his mind." From downstairs the strains of a piano could be heard: it was Jessica practicing her new sonata more deliberately, this time heedful to add further depth and subtlety to each note. "Wait here," Joseph said, getting up gingerly to test the cramps in his abdomen. "I won't be long," he added, before leaving the bedroom.

*

Lexa stretched herself out on the bed, feeling calm and self-assured at having handled Joseph's predicament with such directness and control, with such personal honesty. A gull flew past the windows in the blackened night and its loud screech caused the mice to shuffle in their cages. It was a cosy bedroom. Mowgli jumped up onto the bed and nudged her leg, purring loudly.

Now that Joseph had expressed his fears, now that he had opened up to her, she felt that much closer to him. As far as she knew he hardly spoke to anyone else about the relationship with his father: only seldom did he mention it to Jessica because his sister became depressed. Jessica was still relatively young. Perhaps Joseph had talked to Buck Williams? Lexa wasn't sure. But this thought of his confidence in her left a soft tingling sensation, a gentle flutter of appreciation running down her spine.

Silently Joseph re-entered the bedroom. He switched on the aquarium lamp, then turned off the main lights. The walls of his room and the slanted ceiling with objects hanging off the beams were bathed in a rippling aquamarine glow of water, shadows of tropical cichlid fish danced across Lexa's vision and reflected off a myriad of other darkened shapes in the room. Out of the row of windows now she could make out a few stars glinting through the growing *Kwaaiwater* sea-mist.

Joseph seated himself on the bed next to her.

"Jessica has promised to play us something," he said. He gestured that she remain quiet to listen to it above the muffled pounding of waves outside. Emanating from downstairs, clear, forceful, simple in its beauty, they heard a playing of Beethoven's *Moonlight* sonata. Only once did Jessica play it. In the silence that followed, Lexa was surprised at the tenderness with which Joseph took her hand, kissed the palm with feather-light touches of his lips. He altered his seated posture on the bed so that she could lay her head down on his lap.

"I can trust you with this can't I?" Joseph asked, his voice subdued. "Helping me solve these problems with my father isn't easy."

Lexa looked up.

"I love it when you kiss the palm of my hand like that," she said, smiling. "And that sonata has made my heart race."

Hesitantly Lexa took Joseph's hand and placed it onto her breast. He could feel the rapid beat of her heart. Then unfastening two of the top buttons of her T-shirt, he reached to the warmth beneath. The nipple firmed at his hand's touch. He bent over to kiss her deeply on the mouth, cleaving her lips and receiving the tender, flickering eagerness of her tongue.

Sighing when they withdrew, Lexa asked: "What is happening Joe?" Her voice sounded like an echo in the room. At his continued light nuzzling of her neck beneath the earlobe, she quietened again, enjoying the stroking of her hair against his cheek and the firmness of his hand cupping her breast. It was assurance enough, she felt, that this first time for her was now the right time.

As this nuzzling raised her gooseflesh she slowly, very hesitantly, unbuckled the raw-hide leather belt and drew up the T-shirt over her head. She allowed him to remove her bra himself. Until only the colourful beads around her neck covered her naked torso as she drew Joseph's head to her breast.

"You know it's my first time," she whispered.

Words which mingled so imperceptibly at first for Lexa with her other racing thoughts and the sensations flooding, as to almost go unnoticed. Joseph's mind and stomach which had coiled like a spring with the day's past tensions, with the previous yielding, the tightness in his gut, was now being counteracted by this giving and eagerness of her mouth. Her willing tongue, the pert nipples raised to the touch of his lips. That he felt himself responding in his own way as fast almost as the beat of her heart. It was becoming less of a choice to hold back as Joseph instead tried to forget himself in her infinite softness.

Until those words, her virginity, struck a note. He breathed into her ear: "Are you sure then?"

"Yes!" Lexa's response was immediate.

"I've got protection," he said. "In the bedside drawer."

Her hands seemed aroused with their feminine gentle drawing of nails to raise the gooseflesh on his skin in this aquamarine light, finding their way under his T-shirt too. As she held Joseph to her breast, he could feel himself inwardly uncoiling, unwinding, but slower than her. Lexa began her sensual swim into dizziness as Joseph's lips found her nipples again: she moaned responding to his kiss, arching her back to get closer to his touch.

"Wait..." she stopped him. He looked uncomfortable.

Repositioning herself and bringing Joseph along so that they were both lengthwise on the bed, Lexa lay sideways, her breasts exposed, their weight expressive, their shape voluptuous. He kissed the curve of them. Her legs were equally smooth though firmer and with a more urgent touch of his fingers she drew her leg over his, bending the knee. She allowed his hand to search beneath her skirt where there was a softness that went with whispers of encouragement, her legs parting slightly to the silken underwear. A silky smoothness making him dizzy with impatience.

Their eyes met in acknowledgement that the final barrier was lowered. They both stopped as Lexa got up, kicked off her sandals and removed her skirt. She allowed Joseph to slip off her panties. And in the light of the glowing aquarium, helped him to remove his clothes.

With racy breathing their kisses became even more passionate, as Joseph felt beyond himself begin to trace with his mouth the soft curve of her waist, her hands weaving through his hair. Reaching a place of curls so warm that it made his head spin and her, gasp. She was so beautiful. Together they murmured endearments as they touched completely for the first time in nakedness, their hesitancy, their excitement.

"Oh!" She held breath in pain and relief. Her tongue searched his mouth as he was taken up, closing in on her desires, filling, speeding them with a heat and a motion. Lexa held onto his frame tenderly. Her body was pressing before it was taken up with the tension, away, gripping and feeling the strength of his arms pivoting his weight and the bunched muscles of his back flowing. Her hips moved, stroked, thrust to meet him. It was so new. So new. Until they lost themselves to themselves in this ache of limbs and breath and dreams: in the tumbling deliquescence of it all...

While on the bedside table, unnoticed, silently ticking, the electric clock struck midnight.

- CHAPTER 7 –

Joseph lay awake. Lexa was gone.

In the still grey darkness of early morning his bedroom was a dense configuration of silhouettes bearing down from every angle. From his position lying in bed, he could reach out and touch the mouthparts of the model pterodactyl and imagine himself the terrified prey of some giant swooping predator. There were the pips and squeaks uttered by the caged mice, more frequent at this night-time than during the day. And from his angle of vision the sunbird cage impeded his view out of the rear loft window, barring the night-sky like a prison cell. It was an eerie dark atmosphere that, for anyone in a conflictful state of mind, could transform into dreams of a medieval rat-infested dungeon. Yet Joseph now felt as if he were floating free. Soaring in a thoughtful, peaceful, delicious afterglow.

It had been only his second experience of lovemaking. The first experience of his love life he remembered with some regret. A mistake that happened during his first year at university. Younger and naïve at the time, he fell for the flirtations and flattery of a young lecturer who was quick to recognize Joseph's academic brilliance. On an impulse he accepted her advances. Cajoled by the appreciation of someone academically his superior. It lasted a mere three weeks over the autumn holiday when he had remained behind in the university halls of residence to conduct scientific research with Professor Harris.

But when his classes resumed and he realized her flirtations had suddenly moved elsewhere, he recoiled hurt. Humiliated. He abandoned the naïve chase, allowing his work again to take precedence. He had hardly known her like he knew Lexa. Or ever connected with her mind like he now connected with Lexa. The experience this time round felt very different. Yet this was hardly a subject about which Joseph considered himself much of an expert.

As he lay in the early morning darkness, he realized how unskilled he was in opening up his feelings in the way he avoided telling Lexa about this previous experience. In having lost his virginity in as naive a way as he had lost it. Especially considering how open and honest Lexa had been last night, so truthful and honourable in admitting it was her first time. However, in the heat of the moment, there had been little chance to reveal his past. Joseph guessed she never even realized his lack of experience being so obviously centred herself, he had felt, on her own first time. That he was fully preoccupied in assuaging her own fears of the moment and her own hesitancies. In the heat of the moment, there was little chance for him to share his history. Instead, through whispers of caring and encouragement he had assumed the role of a teacher even though he was hardly qualified.

Although, Joseph thought, she must have realized that I wasn't totally naïve or inexperienced because I offered to use protection. And she was pleased. Appreciated it. Somehow Lexa's sexual inexperience and her confidence in him seemed like a responsibility: at a time of such adult responsibility, could he really have blurted out his inexperience? Such new unexplored questions. A new way of thinking about Lexa. This new dimension to their lives which had unfurled.

Unashamedly Lexa was determined to open this new dimension to their long-term friendship. So prepared, in fact, was she, that in the heat of their decision she admitted - Joseph flinched with surprise at this - to having started on the pill.

"I started a course of oral contraceptives last month", Lexa had said matter-of-factly, as if she had prepared and waited for this moment, as if planned it in silent and determined anticipation. And with her commitment last night and in the responsibility given to Joseph, their excitement heightened only more so. Joseph had never felt himself as ready, as he remembered, now lying alone in his bed. In the end both of them were driven on a response which cares little for inexperience.

Now half awake, released from a great unknown about her and an even greater unknown about himself, Joseph wondered if Lexa was feeling the same. She would surely also be lying awake in bed at home. It was strange how they had never discussed this issue before: they ought to now, perhaps later this morning on the beach. Joseph wondered whether his conversation about the doctoral post had spurred on the encounter. The process of talking about his fears, sharing the conflicts about his father, seemed to draw Lexa closer, even make her bolder. And what a relief to have talked the subject out. To voice the ghost of family guilt. She had listened better than anyone.

Buck, by comparison, knew little about the new options for his future. To him Buck and Lexa were such completely different people. The way Lexa approached the world was to offer feminine practicalities. Even from that first day when Joseph remembered meeting her, she had offered up her own approach following Mr Salem's fishing accident. When the family decided to move permanently to *Kwaaiwater,* towards the end of the year in which Joseph first entered university and before Lexa began her drama studies at Stellenbosch. Often on weekend breaks they would meet up in Hermanus, spending days mapping the marine life along of the shoreline. Or collect specimens for Joseph's taxonomic catalogue. At this time Lexa became more than a friend, more than even a confidant. In knowing his passion for science, surprisingly she was happy to be his fieldwork assistant and librarian too.

By comparison, Buck was less pragmatic, more a philosopher of life. Yet had the old fisherman known the exact state of Joseph's affairs, he may perhaps have been more practical in his own way, more convincing yesterday. Buck's letter remained unopened and would have to wait: that issue of the wandering albatross seemed so far from serious contemplation as Joseph lay infused in thoughts of last night. Languishing beneath the sheets. Sleeping light, immersed in a ferment of playful tingling ideas and sensations. While Mowgli nestled heavily and purred a soft cat's sound in the crook of Joseph's arm.

The night flowed on.

And when eventually the dark sky exchanged its hue for a paler shade of blue, and the stars dimmed, and the pre-dawn prelude of birdsong ushered itself in, Joseph arose on his birthday and took to the bathroom for a cold shower. For an icy, water stinging, breathtaking start to the day that left welts of gooseflesh down his arms and legs. He fed the mice their food pellets, the tropical cichlid fish their pinch of floating food while dressing, and left the house carrying the sunbird in its cage as the first fingers of dawn-light inched over the east horizon.

The *Kwaaiwater* sea breeze was strong and smelled freshly salted. Lines of breakers, four in all building in successive waves, were humming and clattering onto the shore and splitting onto the rocks. Twice Joseph stopped on the winding coastal path: once to allow a Cape robin to hop across the path undisturbed and disappear on its way into shrubbery, and a second time to watch a black butterfly with bright yellow wingtips settle in the dampness beneath sprouting arum lilies. He walked slowly east along the shore as the path curved into the mist. With Danger Point and the caves of *Gansbaai* positioned somewhere beyond the wet haze more than an hour walk away, too far to walk now. Instead, Joseph headed for the long smooth white sands of Grotto beach leaving *Kwaaiwater* behind via the road though *Voelklip.*

Apart from two fishing boats on the distant sea horizon, no one was around to share with him the sight of this morning's African dawn. The path, the gravel roads, the sandy beaches lay deserted of human visitation, but somehow it seemed more spiritual enjoying the sunrise this way. Anyway, after last night Joseph felt a need to be solitary at this moment of his birthday, to wander alone. Though being outdoors with nature how impossible it is to be quite truly alone: far on the seaward horizon, if he wasn't mistaken as he screwed his eyes into the light, an albatross could be seen winging its lonely aerial path lazily this way and that above the Southern Ocean.

On Grotto beach Joseph removed his sandals and sank bare toes into the morning coolness of the sand. The air and sea breeze felt moist with mist, heavy with the odour of seaweed. The sight of the expansive beach stretching into the distance and the sea-taste on his tongue was enough to frisk his blood with energy, enough to spur his muscles on. Putting the sunbird cage down on the beach Joseph headed off in bounding strides, arms outstretched, at first pacing himself for half a kilometer, then running wild into the mist while he could still feel that his skin was sensitive to the salt air from his breathtaking early morning shower.

Shouting: "Yaaahhh... Eeyyaahh!" into the wind, Joseph pounded the beach sand with his furious run. Down along the wet tidal swash as water leapt up to splash his thighs, his hands, his face in the misty haze. Next, he bounded up the beach where it was harder to keep up his speed and pace in the deeper, softer sand. He looked back to check if the bird cage, tiny now, was where he had left it. His breathing was becoming slower, and he began to pant with the exertion with which he egged himself further on, further on, to drain and sap himself deliriously. Until exhausted and gasping for air he crumpled sideways onto a sand dune. Lying flat, head up, chest heaving, while in the distance the albatross continued its flight in slow circles. Joseph watched it mesmerized for a while. Then stood up and walked to the water's edge.

There were infinite things to learn from the sea. He remembered, once, the words of an eminent marine-biologist who was Professor of Zoology at his university. The old professor compared the sea to a woman: "She is a shy and complex mistress," he had once said of the ocean. "A mistress who doesn't easily give up her secrets." Today this mistress, as he had called the sea, was calm and bathed in the yellow tincture of dawn. As waves washed up onto the sand Joseph could see *Bullia* snails tumbling back-and-forth, tumbling back-and-forth in the wave zone of the beach.

Bullia is the name for a type of scavenging whelk, a marine mollusc with an enormous foot. Joseph often watched them rolling up the beach intrigued at the way they extended their foot into a sail to catch the next wave. He called it 'snail surfing'. The way they used this enlarged foot to somersault and thus move up and down the beach: it was quicker that way. *Bullia* had a remarkable ability to smell carrion lying on the shore. A dead fish or a blue-bottle jellyfish or other rotting object washed up attracted *Bullia* within minutes in their multitudes: they would rise to the surface from resting places under the sand and begin rolling towards it.

With interest Joseph followed their somersaulting motion now. As if there may be some kind of lesson to be learned from *Bullia*. Recalling Simon Thandiwe's advice that we should study animals in the wild to learn more about ourselves, more about our own humanity. Without using the waves, Joseph knew that these *Bullia* snails would otherwise hardly ever reach food. Their application of the waves was a bit like Simon Thandiwe finding a way to use the kindness of his community to help him to move towards his political goals.

Inwardly Joseph smiled. Wondering, silently imagining: what if I were a *Bullia?* First I would need a goal and then the method to propel me towards it. He contemplated the doctoral post and Lexa last night; how she had managed to quell those stomach cramps, and how by accepting her strength Joseph felt powered into a decision to confront his father. It was not quite the same as listening to Buck because Buck had advised without really knowing the goal to which Joseph was now entangled. Or did Buck somehow sense it?

Joseph picked up a *Bullia* snail. Its foot was as long as his little finger and not much thicker, its spiral shell coloured pale shades of brown, tan and cream. The creature slid easily across his palm. The animal was able to move fast and unencumbered on the smooth surface of his skin without the abrasiveness of the sand to slow its motion. Joseph contemplated this analogy. As abrasive as the beach, his father too stood as an obstruction before him, barring his movements like an obstacle as wide as this Grotto beach. This was an expanse which had to be navigated. An abrasive, dangerous stretch to be crossed with whatever motion was needed. Glancing across at the miles of sand stretching east Joseph silently cursed having to put the small *Bullia* snail back onto it. But against his sudden flare of anger, Lexa's words of purpose resurfaced like a balm.

Maybe in the end, he thought to himself, maybe I'll also have to choose a way that proves easier. Less encumbered. Perhaps learn how to tumble like *Bullia*, how to somersault myself above and over what, at this moment, looks like an endless, scouring, unmanageable expanse.

*

They had decided to meet at the 'Moby Dick' grave on *Kwaaiwater* beach. Lexa would be waiting at seven o'clock. The time was now six forty-five in the morning. Spinning round Joseph tossed the *Bullia* as far as he could across the moving tumbling waves, silently bestowing the snail a prayer as it skittered over and across the water. Then made his way up the shore collecting the sunbird cage as he went. There was still no other humans in sight although the sun was full, slowly climbing, the albatross now a pin-prick moving its way further and further south.

Feeling vital, expectant-in-love, Joseph strode out. He wondered whether this was what being twenty one was all about? At the road the freedom was his to meander with no people about, swaying one raised arm to mimic that albatross in flight but oblivious of all that Buck had told him. For this precious moment he was no longer the wandering scientist, he was an expectant lover in a lonely and wild place with the spring in his step unstoppable.

Down on *Kwaaiwater* beach beside the 'Moby Dick' grave Lexa was already standing there, facing away from his approach, looking shy and awkward like a gawky schoolgirl, reading the plaque on the wooden 'Moby Dick' cross:

R.I.P.

Here lies what remains of Moby Dick
Forty five feet long and seven feet thick
Once a delight of every whaleen
Now alas deprived of all his baleen
Let us shed a tear in Walker Bay
It makes no difference either way.

"Hello, Lexa."

"Joe..." She turned with sparkling eyes, but hesitated to kiss him until he held out his free hand to her. There was a nervous expectant pause as they smiled. "Do you remember the whale beaching itself?"

Joseph nodded.

"Aren't you glad everyone gave it a decent burial on that day," she said in an attempt to start the conversation. "Oh, you've brought the sunbird!" She noticed, glancing at the cage which dangled loosely in his hand. "Is it ready to be set free?"

He nodded again. "The wing has healed. I'm pleased," he said. Joseph raised the cage as the wild bird in shimmering metallic-green plumage flitted from perch to bars and back. "We can release it later, best up on the *Protea* mountain where we found it. Come let's first go off to the bridge. We can talk there."

She squeezed Joseph's hand as they crossed over the piles of black mussel shells lying scattered on the beach. Reaching the wooden bridge, Lexa sat with her feet dangling over the side above the dark mineral waters of *Mosselrivier*. She beckoned for him to sit.

"Come lean up against me, Joe."

Lexa curled in his grasp. For the first time she could begin to imagine herself sharing more of Joseph than simply his outdoor companionship and his science. He was responding to her. Though she knew she mustn't expect too much, too soon. These things had to work their way slowly.

"I hardly slept last night when I left you," she murmured into the wind.

"Same here." They huddled close as Joseph spoke: "After you went home I lay in bed thinking how much it must have meant to you trusting me like that. It's a real honour being your first time." She sighed as they faced the ocean, breathed in the salty seaweed air.

"I need to tell you this Lexa..." Joseph wavered in these initial words, unsure of how to convey his thoughts to her which he knew he must. In the intervening silence, trying to gather courage, looking away and then back at her, he finally said with some difficulty: "I need to tell you that it wasn't my first experience. But up here," he touched his forehead, "it felt like the first. What happened felt right."

Lexa turned to him. "You don't have to tell me if you don't want to. Other experiences are history." Her eyes searched his face, yet she said nothing further, indeed words didn't seem necessary. The look she gave of attachment to him seemed enough. As she touched his cheek with her hand, ran two fingers to caress his jawline, her head was tilted, her chestnut hair engaging the morning light.

"Have you imagined our future, Joe?"

"What do you mean?"

"No. I guess I shouldn't ask this now."

"Please..." His tone was serious. "After sharing last night, after what has happened, let's try to be open and honest about our feelings. The relationship has changed."

"Then you've thought about it! How it will be different between us."

"Yes, of course."

Her expression went dreamy for a moment. "Last night I lay awake for hours afterwards at home imagining what the future holds."

"Your drama career?" he asked.

"Thinking about that too and your science. But more about us." She looked down at the copper-coloured mineral water trickling into the sand, then up at Joseph's face, then down again. He could see she was struggling to articulate something, her brows knitting themselves in her endeavour to express something.

"Why such a pained look?"

She sighed and looked away. Joseph felt the icy sea spray on his face. He waited as she took a deep breath, building up her confidence.

"I need to know what you want from me!" Lexa said suddenly, gazing back down at the water as he admired her profile strikingly outlined by the bright ocean haze. "It took conviction last night," she continued, "and I need some reassurance that this is going to last."

Joseph hugged her.

"Your feelings are no different to mine," he whispered near to her ear. "It's a big step for me also. For three years we've been very close friends and now things have changed. But it's far too early to make serious plans, don't you think?"

Abruptly she got up. She pulled Joseph to his feet.

"Let's walk," she said.

But before he knew it, she was running ahead. He watched her trim figure moving in her own feminine way up towards the west lookout. Realizing then how much they had to learn from each other, how to be lovers, how to be intimate friends. For so long this intimate moment had been possible: preparing through a mist of fantasy, he had prepared through a mist of facts. Then suddenly the decision - now or never - and they were across the threshold! And somehow, he seemed half cheated realizing he was just a novice to all the emotions set loose. Lexa reached the west lookout standing firm with arms folded, her dress billowing lightly.

She is right, Joseph concluded to himself. Their experience last night meant far more than male fantasy, or female fantasy. He strolled to reach her.

"I think you're right!" Joseph was breathing deeply and rapidly. From this point the westward breadth of Walker Bay lay open before them. She leaned back into his grasp as he put the birdcage down and held his arms tightly around her waist.

"Joe, I need to know that there's place for this kind of commitment in your life. We've never talked about a serious relationship; Jessica or your parents or mine might have jokingly mentioned it in passing, but those were their expectations. Nothing to do with how we feel. And I'm confused: until now our friendship revolved so heavily around your science and your outdoor interests. We've never seriously dated or done anything conventional like that."

"I never stopped you from dating other men."

"Yes, I know Joe. But I wanted you."

"What does that mean?"

Lexa faced him, her hazel-coloured eyes serious. "That it rests more on your shoulders to come to terms with us and your work. Last night I even had a fleeting fear that the pressure of the doctoral post, the farm inheritance, and what with your birthday looming today; well, what happened between us last night might have happened too quickly. You may have been pressurized into it. Perhaps the timing last night wasn't right to get so serious!"

Inwardly Joseph flinched at this statement. However unfair it seemed of her to say this, there was an element of truth, he knew, recognizing that his state of mind last night was reflected in those stomach cramps. Those gut pains were symptomatic of his mental conflict: in all this time he hadn't yet confronted his father about his career. Nonetheless he knew his decision with Lexa last night wasn't made under a callous or needy pretense; it had simply not happened out of any desperate need to redirect or forget about or salve the conflict he was feeling, or to gain any kind of sympathy. He knew quite emphatically it had not happened like that.

"Lexa, believe me! Couldn't you sense last night that I was ready to take this step? Anyway, it was a mutual decision, ours, our decision, not simply mine."

"Then let's discuss where it's taking us."

"We can..." Joseph paused from getting flustered. "But aren't you just a little paranoid for such a quick answer. I mean, where exactly do your expectations or your parents expectations which I know influence you, fit in with these questions? What made YOU decide last night?"

"Oh Joe, it's happened so quickly." She became tearful at their first argument as lovers. "I'm a bit emotional. Just hold me, please." Joseph drew her to him, weaving his fingers through her hair, running them down her neck with repeated strokes as she dropped tears onto his shoulder and onto the lapel of his open shirt.

"This is OUR relationship, Lexa." Joseph searched her face, sweeping her thick hair away from her eyes and wiping the tears. "I'm certain we are both ready for a deeper commitment. Don't you think?"

She nodded. Then she smiled as he proffered a wild yellow daisy: a *Euryops* picked from the ground at their feet, as they both sat down on a chunk of eroded sandstone positioned at a raised vantage point which encompassed the wide vista of the bay.

"Do you need to tell me anything more about your worries for the future?"

"I think I've said what I wanted." She dabbed her eyes with a tissue.

He watched her expectantly.

"Joe, when will you talk to your father?"

"Sometime today."

"He's a generous person - you know - your father is. To have paid your way through university."

Joseph frowned at the context of this.

"Perhaps you could try a bit harder to share with him the fruits of your education. Give him some part to play in your interests."

"What are you getting at, Lexa?"

"I just don't think you should withhold how much you appreciate him!"

Joseph sighed heavily. "Now wait a minute," he said abruptly, his patience teetering. "What about those entrance hall cabinets at home which are filled with my natural history collections. Moved from my bedroom and located there precisely so that Dad can appreciate nature close up from his wheelchair."

"But that's not the point."

"Well what is your point?" Joseph sat stooping forwards on the rock, beginning to dig furrows in the sand at his feet with a lichen-covered stick he had picked up from the path.

"That you find a stable future for yourself."

Joseph stopped digging. He glared at her.

"Now I don't know whether to take that comment as sexist or caring or what exactly?"

"Look, let's just leave it alone Joe," Lexa raised her palms in front of her in submission. "No more arguments please," she implored, "especially not on your birthday."

He shook his head angrily. "But you've just dropped a bombshell. It's only fair to put me in the picture now about what you mean. And neither am I looking for an argument, Lexa, especially around a subject that has been giving me untold aggravation."

She placed her hand on his leg.

"Damn it, Joe. I'm being so vague. There are just all these emotions - incredibly difficult for me to express – pouring through my head. I'm sorry. Promise me though that you'll allow your father to have his say! You know I value his judgements not only because he's older and wiser and more experienced and he's your father, but also because he's a fair judge of character."

Inwardly Joseph cringed at this. However, he refrained from any reply. The furrows he was digging in the sand had reached to the rock beneath and his scraping strokes were making a screeching sound as the stick met the weathered stone.

"Apart from not knowing what your father's reaction will be to the doctoral award, all his comments I've heard about your collections and your farm work have been positive. That should mean something to you."

Lexa watched Joseph's hand moving the stick along the widening sand furrows, though she did not notice how white his clenched knuckles had become or how his fingers were shaking.

"There's nothing worse than fear," Lexa continued. "A playwright I think it was - but I can't quite remember who – once said that fear is the servant of the unknown. I believe that. Now that we are both struggling in our search for some kind of answer to the future, Joe. That decision you make will also affect me."

In silence he stared solidly down into the sand.

"But come on: enough of this seriousness, let's be more optimistic." Playfully she pushed him lightly sideways and Joseph rolled half onto his side on the sandstone boulder. "What other plans do you have for today?"

Joseph straightened up.

"To collect my academic books first, and clothes and things from my room in residence. All science faculty students who have completed their exams must vacate their university halls by the end of today. Stupidly I left the packing up late. On my way back here, I may as well visit Uncle Jack at the farm at *Salemkop*, which means that we'll probably only meet up again later at my birthday party."

Lexa shrugged gracefully and with understanding, leaning against Joseph's shoulder as they got up and began their walk to the *Protea* mountain. She fancied carrying the malachite sunbird in its cage for the last time. Excitedly the bird flittered from perch to bars as if it knew from the sight of the outdoors, the sea and mountain landscape, that it was about to be set free.

Departing from the West Lookout Lexa spoke into the subdued wind: "Thanks for letting me have my say, although I don't know if it came out in the way I wanted it to, Joe. When it comes to discussing issues with your Dad, please try to listen to his advice. Try, if possible, to be as considerate to him as he has always been to you."

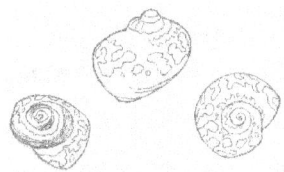

CHAPTER 8

The sun had risen one third of its arc behind the bus as the heavily laden vehicle engaged a higher gear. Forestry plantations of pine stretched outwards on either side of the motorway on which they were travelling, obliterating the view of the nearby *Steenbras* dam. The bus was traversing a high plateau undulating across the Hottentot Holland mountains and making the last ascent on Sir Lowry's Pass. With more straining of the diesel engine, another gear change caused the joint attaching the bus to its trailer to jolt and jerk with a loud "CLACK-CLACK". The trailer being pulled behind was almost a quarter the size of the bus, painted in the same colours of red and grey. It contained luggage, parcel post, frozen meat, anything needed to be transported back and forth from city to small town. This was the Bredasdorp-Hermanus bus line, their destination the South African Airways courtyard in the city of Cape Town.

Joseph sat three rows behind the bus driver, leaning against a large rectangular window. Pine trees flashed by in geometric rows as he looked out onto the lush green scenery of these forestry plantations. Joseph was unable, however, to see the trailer being pulled by the bus but then neither could any of the other *white* passengers, and there weren't many *whites* (or 'first class' customers) travelling today. That sight of the trailer was reserved for *'non-white'* eyes only. Three rows of seats at the front of the bus were reserved for *whites* and nine rows at the back for *non-whites*, completely separated and sealed off by a metal partition. Thus, there were two entrances to the bus: back and front. Two social classes to the bus, two skin colours, and from the singing which Joseph could hear behind the metal partition it seemed even two degrees of happiness to the bus. Perhaps *whites*, he wondered, felt more secure sitting at the front so that they arrived at their destination first, always keeping a little ahead. Joseph rested his cheek onto the metal partition and could hear behind it a happy raucous: numerous hands clapping, songs of an ethnic origin, the stamping of feet. It was a carefree and quite joyful African sound that at this quiet moment he wished he could be a part of.

The journey to Cape Town was about two and a quarter hours on the bus. There had always been a route over these rugged mountains: the wild eland antelope were one of the first to find it in their age-old search for better pastures; and later came the Hottentot people driving their cattle. These Hottentots followed the eland over the *kloof* and called it *Gantouw* - the Eland's Path. And in their footsteps came the early Dutch settlers, but the Eland's Path was too steep for them. At first everything was carried up the *kloof* on pack oxen with even their wagons being reassembled at the top. Later, these hardy Afrikaner farmers evolved a route up *Gantouw* and for 150 years this Hottentots Holland *kloof* was to remain the chief highway to the east. By 1828 so many of the 4500 ox-wagons using the *kloof* each year were damaged that the British governor, Sir Lowry Cole, saw to the construction of a proper graded road across the face of the mountain. At the opening in 1830, it was decided to name it Sir Lowry's Pass.

The bus surmounted the summit. A hush descended behind the metal partition and Joseph could hear passengers crowding to the left-hand-side windows. The view over the high rocky edge, a panorama as much scenic as it was expansive, was breathtaking as if the bus had suddenly taken flight. The road clung to the sheer rock massif as the land beneath it poured down in a steep cliff-face and then out flat into False Bay and up to Table Mountain more than 50 kilometers away. Beneath them the peninsula lay naked in the ravishing beauty of its winelands. Cultivated vineyards a hundred different shades of green, striped with neat rows of grapevines. A

191

sight of paradise for the mathematician and the geometrist: squares, rectangles, triangles, fringed with pine tree hedges. Reservoirs and farm dams were dotted around and as they caught the sunlight they glistened in spangles of silver; homesteads resembled toys nestling amongst their toy gardens; whilst the Indian Ocean stretched out wide to the left in an almost endless emerald arc.

The bus descended Sir Lowry's Pass with a gathering speed. Soon it was driving steady along the Cape Flats and the motorway scenery became suburban. A middle-aged couple whom Joseph did not recognize from Hermanus were the only other two passengers in this *white's only* part of the bus. For the past half hour Joseph had been reading his birthday gift from Lexa: Konrad Lorenz's book *On Aggression*, but became distracted by another hush from behind the metal partition and his concentration moved from book back to window.

They were traversing the outer perimeter of the large shanty townships of Old Crossroads, Nyanga Extension and KTC. The double lane motorway was reduced to a single left-hand lane by a cordon of oil drums. Armoured vehicles of the South African Defence Force were patrolling the lane closed by these obstructions: between abandoned and damaged cars which had been stoned, and barricades of smouldering tyres. The road was littered with a confetti of stones, bricks, tin cans and broken bottles. Thin columns of smoke were rising darkly over the carcass of a neighbourhood of burnt out shanty houses. Incredibly, a few black residents in their everyday need to continue carrying out their domestic chores were hanging recently washed clothes on the surrounding rolls of coiled barbed wire. Piles of family possessions rescued from the flames, which included furniture and mattresses, lay dumped and scattered on pavements.

A whiff of smoke suddenly penetrated the windows of the bus. It made Joseph's eyes sting and water and burn. It was an unexpected thick gust of toxic black vapour which smelled acidic and pungent, a heady mixture of smouldering rubber tyres and the burnt odour from charcoaled shanties. Joseph thought he smelled tear gas in it too which was likely. Then suddenly from behind the metal partition of the bus, a roar of shouting commenced from the *non-white* passengers as an armoured vehicle manned by *white* soldiers was passed.

"GET OUT OF CROSSROADS!" some passengers shouted in unison, while others chanted the phrase over and over.

"WHITE APARTHEID ARMY GET OUT OF CROSSROADS!" a few male voices roared.

"S.A.D.F. VOETSAK! S.A.D.F ARMY GO HOME!"

The elderly couple sitting adjacent to Joseph in this front part of the bus glanced at him with an expression of alarm. They huddled together, husband and wife in all probability, on a bench seat. Joseph acknowledged them with a hesitant smile that revealed his own alarm. Then he turned to continue looking out of the window: as the soldiers were passed, dwindling in size into the distance, the voices behind the metal partition quietened into a vacuous silence which seemed desolate, devoid of their previous happy singing.

The odour of smoke remained strong in the bus, powerfully acrid and unpleasant, although the haze in the air had cleared. Then as another armoured vehicle was approached, loud protestations resumed and fists were held out of the bus windows. The noise attracted a dog outside who began to run beside the bus barking excitedly at the people chanting inside. Until the bus accelerated when the two lanes of motorway opened up again - free of the oil drum cordon - and the dog was left trailing behind, still mouthing soundless barks and wagging its tail, its run blocked off from further view of Joseph's eyes by the metal partition inside the bus.

The elderly couple sitting near him, Joseph could plainly see, were shaking in fear now whilst still huddled tightly together. Joseph examined his own hands for any nervousness while holding them out and noticed that his fingers, wrists and arms were quite stationary, calm and still. He felt oddly empowered by the experience he had just witnessed. He felt a clarity of mind. As if Simon Thandiwe were sitting here reassuring him. "These are the underdogs," Joseph could almost hear him say pointing outside. "You have seen my people fighting back. Ordinary people…" he imagined Simon would say. In the way this place was swarming with the South African military, in the spectacle of township devastation, a war zone, Joseph could understand why Simon Thandiwe had had to escape it and flee to Hermanus.

He also remembered the simplicity of Simon's words yesterday in the presence of Buck. "Try to respect both sides". How Simon had equated this respect also to Joseph's studies in biology, to the better understanding of animals, as well as resolving the conflict with his father. In the calmness and crystal clarity of his own mind now, Joseph began to feel an understanding of how Simon could have been so calmly spoken, so self-assured in this knowledge, quite resolute that he would succeed even while all around him burned and smouldered.

In the book *On Aggression* which Joseph picked up from his lap and continued to read, Konrad Lorenz, the great Nobel Prize winning biologist had written about the merits of controlled aggression!   And the value of interpreting behaviour in the way in which it has evolved. Seventeen pages into the book and Joseph began a sudden fervent search for the possible relevance of this writing to his predicament. Perhaps even its relevance to the apartheid war outside the bus windows. While behind the metal partition of the bus, the other passengers had resumed their laughter and African songs. For a few minutes Joseph listened to them in their gaiety as the bus sped past makeshift plastic homes of squatters who had fled from the carnage some kilometers back.

In his book Konrad Lorenz was debating the concept: 'The struggle for existence' not as the struggle between different species so much as the competition between near relations, near relatives, which Darwin believed to be the principal force that drives evolution. It was 'competition' between those nearest to us, those of our own species, other humans. And in this everlasting gamble of hereditary change, a profitable invention or trait may befall by chance one or a few members of the species. Only 'profitable' traits will succeed in this competition between near relations.

Joseph wondered if a gift for biology could be considered a profitable trait? The problem contemplating evolution in this way was the aspect of human culture. It was a subject not fully integrated into either Darwin's or Lorenz's theories. With human culture it was questionable whether any of this Lorenzian or Darwinian thinking had relevance for a trait or a talent that by all accounts was expressed culturally. Because the most ideal human cultures which espouse equal rights, equal opportunity, individual choice, allow the harsh 'struggle for existence' to be overcome for humans in evolutionary terms. In the ideal human society, individuals can choose their fate in life or be assisted by society. In these terms, unfortunately being a talented scientist seemed no less profitable than being a successful wine farmer. Without a culture of individual tolerance and acceptance, when you are ruled by dogmatic assertions of those simply wanting to dominate, like, Joseph felt, the assertions of his father or the war of intolerance outside the bus, this animal 'struggle for existence' may become an element of truth for humans.

On the next page Lorenz described a form of fighting behaviour: a scenario called the critical reaction, a chilling scenario which included the term 'fighting like a cornered rat'. Lorenz wrote of it being a desperate struggle in which the fighter under threat stakes all because it cannot escape and can expect no mercy. This is a violent form of fighting behaviour motivated by fear, by intense flight impulses whose natural outlet is prevented because the danger is too near; so the animal not daring to turn its back, fights for its life with the proverbial courage of desperation.

These threats as Lorenz described were not necessarily only physical enemies. They could be threats of expectation. Future threats. Threats of rejection. Threats of alienation. While the human mind, Joseph was aware through his research on consciousness, is the most well developed and can be the most dangerous adversary of all. In being able to coldly accept such injustices as they had just passed on the bus. Like a boxer being stalked in the night by its own shadow, that shadow of the mind cannot merely be grabbed or wrung out like apartheid or by unjust family expectations. Those shadows can just as easily glide into the very depths of your fear. And people can accept them, are affected by them, often not even knowing why. It was like his father's words on the day of the fishing accident, unforgettable images bobbing in the sea, especially the memory of his father's delirious incantations:

"Joe. That you, Joe! Please God, don't leave me... son, don't ever leave me...!"

Words which hovered above Joseph's mind like a sword of Damocles. They still affected him deeply. These words which he knew he had to begin to fight against. For the years of remembering this incantation uttered in the waves, Joseph had felt cornered like a rat confronted by a rattlesnake, like that poster in his bedroom of the prodigal baboon facing the onslaught of the leopard on a lonely desert salt pan. Yet if he were to see these words for what they were, both the rattlesnake and leopard dissolved into human enemies: peer pressures, parental pressures, apartheid pressures, and one's very own compromising will to do what others want or expect of you. Where is the truth or heart in that? The heart which speaks of self-fulfillment, self-respect, adulthood? If what others want of you conflicts with the driving urgency of your own talents!

Even the wisdom which Buck conveyed in knowing Joseph since he was a young boy had made it sound too simple.

*

Compared to his cluttered laboratory bedroom at *Kwaaiwater,* the decor in Joseph's university room in residence was austere, almost monastically so. The room was far smaller in size, paneled in dark oak, and distinguished more by the lack of interesting clutter than the presence of any collections of natural history or live plants or animals. This room had merely served Joseph as a pit stop in his days of intensive academic study on campus.

Some of the contents of his wardrobe had already been transported back home to Hermanus on Thursday afternoon. Now there were mainly textbooks and lever-arch files to remove, a series of science dictionaries, thesaurus, a wad of postcards tied with an elastic band and sent from overseas universities requesting reprints of his 'Whip Scorpion' article, aside from the photographs and reference notes off the pin-board. From the wall he detached two large landscape posters and rolled them up: one of Cataract Canyon in Utah, and the other of a lone territorial bull wildebeest standing guard in the thornveld at Etosha National Park. These things and a few last items of clothes and oddments would all fit easily into a cardboard box, rucksack, and a shoulder bag.

"Happy Birthday mate."

"Tim! You're still here on campus. I thought you'd left for Zimbabwe last night?"

"John cancelled flying at the last moment so we were forced to find replacement transport, which is leaving midday today." Tim looked at his wristwatch. "And you, did you travel back here by coach?"

Joseph nodded. "I'm collecting the last of my stuff. Have you seen Professor Harris today?"

"I did earlier this morning, walking across the science faculty. But why on earth do you want to see him? C'mon Joe, take a break, varsity's just broken up!"

"It's to do with my residence fees, Tim," Joseph lied. "That's all."

Tim knew nothing about the doctoral award because Professor Harris had urged privacy until Joseph formally accepted the post. Effectively a news blackout. To Joseph it seemed a harmless enough lie to spin about not paying his fees as Professor Harris was also warden of their residence. Following the Hermanus bus trip, on the short train ride through Cape Town to campus between watching the passing urban scenery of Saltriver, Observatory, Mowbray, Rosebank, Joseph had contemplated whether he should discuss his family predicament with Professor Harris. Now he settled his thoughts straight in preparation for such a talk: late morning would be the best time to see the busy professor. It took Joseph a thoughtful few minutes to pack up his books and other things, as Tim rattled on endlessly about a forthcoming Friday night football match.

*

"Please come in," Professor Harris's daughter answered the door, her manner of introduction efficient, practiced, in the way she invited Joseph into the Wardens residence apartment. She was in her second year of undergraduate studies, tomboyish in appearance, with cropped brown hair and dressed casually in jeans and a logo T-shirt. Joseph knew of her high academic reputation on campus. She led him through an entrance hall lined with rather formal looking oil-painted portraits of previous wardens of this residence, before passing into a reception living room.

"My father is busy at the moment but said he won't be very long. You can take a seat here." By appearance alone she seemed an academic through and through. Joseph guessed that, with a father who was both professor and warden, she had probably spent her entire childhood growing up around university students.

Joseph sat and waited, glancing around the formal room, stark with dark wooden furnishings, not particularly welcoming in its atmosphere. A few ornaments were arranged with almost scientific precision on three small coffee tables and inside a glass-fronted teak cabinet. Above the mantelpiece, the focal point of the room was established by an imposing painting of Charles Darwin which added a severe Victorian touch to the decor. Darwin's eyes were intense and dark, brooding and thrown into shadow, with a black overcoat draped across the bearded old man's shoulders. Joseph got up and approached the painting, looking up at the figure of the great scientist with a sense of respect and awe.

It seemed fitting this painting was hanging here since Professor Harris himself was a biologist. Though the room was so surprisingly bland and uninteresting otherwise, as if the great man Darwin presided as an attraction in the absence of any of the profuse nature which he so boundlessly and vividly had described. The room expressed nothing of that. To Joseph a fitting epitaph for Darwin would be to have a room packed full of specimens of nature or art or photographs depicting nature showing its infinite array of colours, weirdness and wonders.

Painted sitting beside Darwin in this imposing portrait was one of his young grandchildren. Darwin was a confirmed family man. Suddenly, to Joseph, that seemed important. Charles Darwin did not live the life of a lone and maverick scientist, a wandering albatross. In mind, indeed, perhaps he was! And in his early years he certainly did wander the world on his voyage of discovery sailing onboard 'The Beagle'. But the image of Darwin the intrepid explorer was soon replaced with marriage and family, a sober life of work and fathering, of Victorian relaxation. A mentor of such stature as Darwin quite clearly made nonsense of the wandering albatross image. But Darwin was a man of independent inherited wealth who luckily had considerable family support.

Professor Harris entered the room. His face registered some surprise that the person waiting for him was Joseph Salem, one of his exceptional postgraduate students. Unobtrusive by his entrance, the professor stopped to silently watch Joseph who remained oblivious of him and quite preoccupied with an owlish stare at the painting of Darwin's portrait. To the professor it seemed unusual to encounter Joseph this self-absorbed and inattentive of someone else which reminded the man of that strange avoidance behaviour in the library on Thursday.

"A-hh, Joseph!" he finally uttered.

It took Joseph some moments to realize another person had entered the reception room.

"Professor Harris..." Joseph fumbled his greeting. "Good morning, Sir. I'm glad you managed to spare some time to meet me."

"Please make yourself comfortable." He gestured that Joseph sit down beside a circular embossed coffee table and himself pulled up a burgundy leather upholstered chair. "How's the packing going? Most men should have vacated their rooms by now."

"I'm done. And West Annex seems empty."

"Good." The professor paused, crossing one leg over the other. He was a tall distinguished-looking man, soft grey hair, genial and open in conversation. "Anything that I can help you with, Joseph?"

"I wanted to discuss the doctoral post."

"Are you undecided?"

"Myself not, no!" Joseph sat back calmly. "But there are certain problems I may have at home."

Professor Harris raised an eyebrow to this.

"Go on," he prompted. "Indulge me."

Since unburdening himself last night in Lexa's company all of Joseph's concerns and fears flowed easily, unemotionally, without him having to mention the guilt or morbid feelings of despondency. Professor Harris, who had never met Mr Salem, listened with keen interest: about Joseph's 21st birthday, the farm inheritance, his father's expectations. The story wasn't that unusual although the argument did have its unique points. What was unique was Joseph's unflinching dedication to biology, uncompromising to the point of holding firmly against a father who apparently had ruthless, unblinking dreams for his son.

Professor Harris nodded attentively. It wasn't the first time he had seen the dreams of a student snuffed out, in fact it happened often enough: it was an understated fact how many talented young adults compromised crucial years of their life acting out their parents or someone else's will, never really realizing their own unique potential and dreamily repeating this cycle of hope and expectation on their own children. At ten-year reunion celebrations, he often met these past students.

"Firstly Joseph, happy birthday!" the professor began. "Now it would indeed be unwise for me to take sides in this matter. For one I don't consider myself an impartial observer since I would very much like to see you accept the doctoral post. Also, someone like me ought not to meddle in another family's affairs. I, myself, as a father, have my own fears and uncertainties for my children, as to whether my daughter and my younger son are well placed in the sciences or the humanities, or in some other discipline perhaps, and dare not over-influence them."

"I didn't realize that," said Joseph.

"Liz - my daughter who opened the door to you – has already made that choice in physics and although, I am told, she shows exceptional promise, it is clear to both my wife and myself how unconventional the rest of her life is."

"I've heard she's very bright, Sir, and dedicated. Which I must say I do admire."

"So do I, Joseph, as I admire the quite unusual talent you show in your studies. However, sometimes it's difficult as a father having to watch my daughter be so hard driven as to forfeit important aspects of her emotional growth. Her social life. The enjoyment of normal everyday interactions with people. At her age I wasn't nearly as dedicated to do that, but on the other hand I don't believe I possessed the measure of talent which my daughter is blessed with."

For an instant the professor struck a sad fatherly expression.

"But you know..." he said, pausing in thought. The distinguished man uncrossed his legs and leaned forward openly, in confidence. "A common failing among dedicated science academics is that they often shy away from the outside world. From normal everyday life. The hardships off campus which other people have no choice but to face. Some academics I've known are quite unable to cope properly outside of university life. I have personal thoughts on this: on how science and technology have come to be used and abused because of this imbalance or weakness even of scientists to engage with the real world outside of their particular subject. But perhaps that is a digression?"

"No, go on, Prof. I'm fascinated."

"My feeling is that on a one-to-one level, Joseph, scientists seldom stand their ground when it matters with those in power such as financiers or politicians. They shy away from confrontation, from the strength that is needed to stand up to people who know less about science than they do, but who can wield the power of science. If you examine the history of scientists, their talents, or their discoveries, more often than not their work has been used like a pawn in the human power game. Over there is an example..."

The professor extended an open hand to the portrait of Charles Darwin above the mantlepiece.

"It's debatable whether Darwin's theories would ever have seen the proper light of day were it not for the fine and brave fight put up by his colleague Thomas Huxley. They called Huxley: 'Darwin's bulldog'. Huxley literally had to stand up and ward off the public disgracing and manipulation of Darwin's theories. Because Darwin himself shied away from public confrontation; at first, he was afraid even to publish his views in his own lifetime and after his book *On the Origin of Species* was published, he became as reclusive as that shadowy portrait illustrates."

Joseph looked up at the painting. He had never conceived of Darwin - his hero and mentor - in this way. That image of a lone albatross in Buck's words actually fitted Darwin.

"Prof, do you think scientists are any different now to then?"

He shrugged. "Perhaps not. But we live in a completely different era. Hopefully we've had time to learn from our mistakes. Establish ways of getting our views more forcefully across. In this day and age scientists should understand a great deal more of the consequences of human actions than ever they did. I also believe that, in

207

being specialists, scientists should appreciate more the abuse of human power on the subjects they know most about. Yet for example – as biologists - we still seem incapable of being effective in persuading those in power about something as basic, as dire in its urgency, as the mass extinction of plants and animals that is happening today."

"It is one area, Sir, where I hope I may be able to contribute my efforts."

"As a scientist, Joseph, and as a South African, I would guess that you have not blinkered yourself to this era in which we live. As a biologist here at university we have taught you the importance of conserving nature. The tenet of ecology. But we have also taught you to question what you observe around you. So as a South African - and I ought not to take political sides - you must have seen on the warring streets of our cities how lamentably wrong and short-sighted politicians or people in positions of power can be. Joseph you're in a unique position today, even on your birthday, especially on your birthday, to weigh up your conscience to these considerations."

"It is some responsibility."

The professor smiled. "But having the talents you have as a scientist *is* some responsibility!" He sat back in his seat. "What I'm getting to in a quite candid way, Joseph, because I trust you, I trust your judgement, is the need for self-integrity. Nothing more. Nothing less. Stand up for what you believe no matter the threats around you. If you face this wisely, thoughtfully, I honestly believe you will make the correct decision. And don't forget to listen to your emotions: no matter what other scientists say, because inner emotions, and inner doubts, serve to your advantage. Don't shy away from them."

It was as if he could see the uncertainties swimming in Joseph's eyes. Once again the professor himself manifested a momentary sadness, an expression of nostalgia in his own eyes, over and above his status as a father. He shrugged the nostalgia away.

"Hah," he uttered, "it brings back memories. I really envy you this moment Joseph. Having reached this crossroads so early in your life."

"How do you mean?"

"Being able to consider where your responsibilities lie in the way you are now doing. Many people, many scientists included, never do consider the depth and breadth of such responsibilities their whole lives. Or if they do, when they eventually do, mostly later on in life, seldom are they free enough or energetic or inspired enough anymore to do much about it."

"I'm afraid you've lost me, Sir."

The professor crossed his legs. He smiled, contemplating the degree of innocence on Joseph's face. The natural and inexorable way in which Joseph seemed eager to tackle such fundamental questions not only about science, but about intensely personal and even wider social priorities he was aware of in the country around him, without even realizing the wisdom of his approach.

"I envy you Joseph. Unlike Liz, my daughter, you aren't simply just channeling or narrowing yourself into science - but more you're broadening yourself into it. You question your father's motives in your career which is quite right. Quite valid. But I'm more impressed that in this questioning you remain so sensitive and you take such pains as to what your father might feel or think."

"I suppose I do care a lot."

"At this moment we're having a quiet and reserved conversation, Joseph. But you won't believe it if I say that right now I feel like standing up, pulling you to your feet, holding your arm up high and shouting out loud: DON'T SWERVE UNTIL YOU HAVE IT!"

Joseph burst out laughing.

The professor's face showed a tenderness, almost of fatherly concern.

"No doubt your decision involves many people other than yourself. But this doctoral award is an opportunity to personally discover *your* talents, *your* dreams, to give them a chance to flow and run while you still have relatively few responsibilities. Joseph, I know this decision and the conflicts involved in them must seem harsh to you. But once you've made up your mind, once your choice is final and you're more at peace with yourself, in time there'll be many other responsibilities you'll have to consider and this will put you in good stead. Very few young adults make use of this important individual choice while they're still free and unattached, so few use it wisely in making the big choice at this early stage of their lives in order to propel themselves onto the level of their own talents."

The professor scanned the room slowly. "I still sometimes wonder," he continued, "about what would have happened to my career had I not started a family so early. My dedication to research, the long hard hours and intense periods of study obviously had to be compromised. It is only right under those family circumstances that one's efforts are diluted. Personal goals I set myself at age 30 could not be reached, as I saw them falling away at ages 40, 45 and 50. It's funny you know, but I doggedly set myself book writing targets which I was quite clear about. Yet each time my wife and I smoothed over the career compromises with other thoughts: family, home, domestic involvements, which were different but not equal replacements nor could they ever be. Because - Joseph - at the heart of it I still envisage myself as the researcher in search of that special scientific discovery which, had I been unswerving those years ago, I know I could have achieved. But now I'm weighed down with administrative duties: Warden of Residence, Head of the Zoology Department. That's why I urge you on Joseph, I urge you: NOT TO SWERVE UNTIL YOU HAVE IT!"

Professor Harris's clear intelligent eyes remained unyielding on the young student in front of him. They were compassionate eyes.

"Takes me back to that day when I had to make the same decision as you. "It wasn't easy... " There was an almost imperceptible quaver in the professor's voice, a quaver of sadness. "I hope my advice means something to you."

"It does, Sir."

Joseph glanced at the elegant grey-haired professor who, unusual in his nostalgia and apparently not convinced himself, seemed to have so much: a high degree of success as a scientist, good sense and integrity, prestige in the science community. He had a stream of publications to his name with the concomitant recognition and status that it brought. A stable family life. Joseph knew little about Professor Harris's past: merely that he was from a family of academics and his father had been a noted civil engineer. Surely then, Joseph wondered, the professor would not have met with the kind of parental conflict that he himself was facing. What exactly did the professor mean when he said of his past decision:

"It wasn't easy?"

Liz, his daughter, was in science. She had established a reputation as one of the brightest physics students on campus. Perhaps she didn't realize how fortunate she was to have a father who seemed to understand her dedication and could live with it.

"I could talk with your parents, Joseph, but I imagine it's more of a personal family issue."

Without hesitation Joseph held up his hand and nodded in agreement. The last thing he wanted was to involve Professor Harris in an unpleasant domestic tussle with his father. It was quite possible, Joseph knew, that his father would compel the professor to travel all the way to *Kwaaiwater* to initiate a fraught doctoral confrontation.

"No, this is definitely my decision alone," Joseph confirmed. "Of that I'm certain."

"Joseph, promise me that you won't short-change your dreams. You've worked so hard for them. If I thought you were just another average student, you would never have been offered the doctoral post in the first place. Think about that too. Only very few students, very few scientists are blessed with a capacity to generate brand new ideas. You, Joseph, are so blessed."

As appealing as was the compassion in the professor's sentiments and in his eyes, it struck Joseph at last that this desperate search in himself for an answer couldn't just be a search for approval from any of these people: neither from Buck, nor Lexa, nor Professor Harris. Let alone his father. Those who really cared, Joseph reminded himself, would come to accept his choice. Professor Harris's daughter had already made her choice.

So had Buck chosen his path, and Simon Thandiwe, over and above what other people wanted or perhaps expected of them. As the tall and distinguished professor sat silently sucking on a cigar, Joseph thought on his words:

"It wasn't easy."

And remembered Professor Harris's ecstatic response when on that day not long ago he had announced the results of his desert chameleon project. It was as though a spangle of light, a brilliant ray of sunshine, had crept into the professor's office and freed him from the mounds of administrative paperwork. Why hadn't Joseph seen Professor Harris that excited more often? How could the man's good sense, judgement, and his integrity deny him that satisfaction? How could they deny him the expression of his own considerable talents for science? Unless Professor Harris, in his own nostalgic sadness, hadn't really made the right choice for himself those many years ago!

Joseph stood up and was on his way.

- CHAPTER 9 -

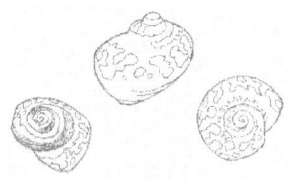

Thin pencilled strokes of sunlight, beams of long and hazy illumination left a pattern of mottled shade on the ground and upon the surface of the river. Afternoon light was penetrating the dense tree canopy of the riverbank, passing between leaves and branches and casting shafts of brightness into the open spaces, furnacing the air beneath the drooping weeping willows. A sunlight which played illusory light games on the water of the river whose surface slowly rippled as it flowed past and on the clouds of swarming midges: fine black clouds of insects in their thousands which rotated above the water in a hum of peaceful circles.

Dear *Muishond-kind,*

You frighten me with your words of sharp accusation. Please understand that I did not judge what you should do or what you should be. It is unfair to say so, it hurts me deep in my chest to hear this folly.

My *seuntjie,* we both have a great deal to learn from that song 'born under a wandering star'. I speak through many years of rejection as a *coloured* man, years which have broken the hearts and backs of many of my people. You see how angry they are now because in this country we have developed the wrong philosophy: blind dependency...

The *white* man has come to depend on us, the *non-whites*, for his comforts. Reliant on us to pull gold out the ground for him, look after his children, cook his meals and make his bed.

And because of this the white man has developed a false sense of pride in himself. I see a more genuine pride amongst my people, not because they have next-to-nothing but simply because what they have they hold with dignity. With legitimacy. No guilt attached.

Whites enclose themselves in so many lies, fit blinkers to the many reminders in the streets of this blind dependency that is not a positive kind of dependency. But a greedy one. And it's not only a physical thing this child-like dependency that treats people like slaves, it's also of the mind: oppression, that double-edged sword, is also psychological and in the process the *whites* have cornered themselves.

And please my *seuntjie*, don't get me wrong, don't get us wrong, it's not that we want to steal those *white* possessions. Often my people don't even like those high-falu'tin material things anyway. It's just that we also need the

216

opportunity to be able wander around freely like an albatross; it has helped me and it will surely help you.

When a man can say: "I am my own free man!" and a woman, child, a society can say: "We are proud of ourselves because what we have is our own!" then we'll have reached the goal of our liberation. Because my son, if what we have is truly our own, we will want to share it.

Ever yours,

Buck.

Joseph sat with eyes closed, at rest while leaning up against a willow trunk, feeling tranquil in this place as he listened to the soft gurgle of the river. He folded Buck's letter thoughtfully and put it back into his shirt pocket. He had been lucky to be given a lift to *Salemkop* farm by Tim who was heading on a long outbound journey to the city of Johannesburg and then up to Harare in Zimbabwe. And Joseph decided to travel with him as far as the Paarl wine district to enter the farm from the north-west side, legging it over a wire fence into the property and through an open pasture, then rock-hopping across the shallows of the river.

In this shade of mottled light there were a myriad of insect sounds and in fact as many movements as there were sounds. The communal hum of midges as the swarms spun their lazy circles over the water. The 'ZZZZTTT... ZZZTTTT...' stop-start darting motion of dragonflies and the quick flight of more delicate damselflies. Over near the water's edge by the riverbank where Joseph was resting, a cluster of noisy blue-flies were in a frenzy of buzzing while making feast of some soft cow dropping. A cool breeze sent a shiver of movement along the top of the willow canopy, swishing the branches and causing a few leaves to sprinkle and cascade to settle as a crisp green confetti on the ground.

Joseph opened his eyes and turned his head in the direction of a rapidly repeated birdsong: 'A-ZWIT... A-ZWIT... A-ZWIT'. The song which he recognized belonged to a Cape weaver bird that could be seen hanging upside down off its nest attached to a drooping willow branch. It was a male in full breeding plumage: body a greenish-yellow, bill black and heavy, head a red-brown, nape and sides of the face a light saffron. 'A-ZWIT... A-ZWIT...' it called again and again. Although there were a dozen other kidney-shaped nests nearby, this was the only male in sight, fluttering its wings and calling for a female to come to inspect its nest work: 'A-ZWIT... A-ZWIT...'

Joseph closed his eyes again. The familiarity of this place, its sounds and its smells, had an intoxicating quality. Joseph unlaced and removed his boots then his socks, to sink his toes beneath the carpet of willow leaves deep into the cool humus. He did the same with his hands while groping around in the leaf litter, gently feeling the squirm and coiling of earthworms, the leaves crackling and breaking in their dryness. With his fingers and toes buried at this ground level at which ants and beetles were active, his vision could focus down the slope onto a thousand emerging grass-stalks. It was almost as if his limbs were plugged into some earthy battery - into tremors of his past, of his childhood spent down by this river exploring, observing, inventing things as a boy. He felt much the same person then as now.

He could visualize, behind his closed eyelids, the old treehouse which had once perched a third the way up a giant willow tree that still stood on the opposite bank of the river. The treehouse was no longer there. Although on the tree remained a few nailed planks of wood and the fork in which the treehouse had nestled. Joseph recalled how he had often used the treehouse as a hide from which to observe the Cape weaver birds. With boyish fervour and relying on the accuracy of a telescope set on a tripod, he had recorded in his youthful handwriting each phase of the lifecycle of these birds: incubation and birth, chick-rearing, adolescence, courting and mating.

On a hot summer day just like this one he had watched horrified yet enthralled as a mother bird, oblivious of all but her duties to her eggs, was taken by a tree snake without so much as a murmur of alarm. It was the type of experience that you want to lash out at and do something to help: jump from the tree into the river, shouting, splashing the water, waving your arms, throwing sticks and sand at the *boomslang* snake - but Joseph recalled how his actions froze at the time. He had stopped any human intervention. To instead watch the unfolding tragedy in awe. Because it was nature's call. And only by leaving nature alone to run its often harsh and brutal course could he begin to learn its ways. The experience taught him how difficult it is to be an impartial observer.

That whole summer he had spent observing these riverside birds. In fact, he knew the Cape weaver so well he could now guess at the fate of that singing male. Its call seemed too impressive - the bird looked too full of itself - it was probably still an inexperienced sub-adult. Certainly, the bird would attract for itself a female as males in breeding plumage most often did. Joseph listened as it continued to call ecstatically, perched on a branch adjacent to its nest: "A-ZWIT... A-ZWIT... A-ZWIT...". The nest, however, even from this distance was a mess. Its kidney-shape was skew in construction and the opening was incorrectly angled. It was the elaborate structure of this nest that was the object of choice – whether to accept or reject the male - by the dull-coloured female.

First, she would be patiently enticed. The male would take great pains to court her and attract her to test its entrance perch and if he was lucky, she would disappear into the little abode. But as Joseph had seen often enough in nests so botched and incompetently made as this one, she would reject it with utter ruthlessness by poking her beak straight through the walls. Her standards of workmanship demanded only the best. The finest of nests. And if he gave less, she would dismantle his hours of difficult construction work without mercy: the many strands of intricately knotted grass would be shredded in a frenzy, bits of delicately woven leaves would be pulled out with ferocious biting action and cast aside. Finally as a *coupe de grace* she would often drop the whole sorry construction as if with female contempt into the flowing river. In doing so, with a cursory flutter of wings, she would reject the male and force him to begin again.

Even the pattern of riffles on the water hadn't changed much since his boyhood. This mountain river was still icy to the touch, tinted dark in colour by the mountain minerals and healthy for both swimming and drinking. In shadowy places between the reeds and between aquatic plants and the submerged willow trunks, under the banks and the riverine rocks, was a whole ecosystem of life. Tadpoles and fish were there to be found as were crabs and dragonfly larvae. Joseph watched as a dozen aquatic beetles – named water boatmen for their sailing skills - sliced gracefully across the surface as if taking part in some ice-ballet extravaganza. He knew that if you took an eye-dropper and sucked up some of the mud from the riverbank, as he had often done and then place it under a microscope, you would find there a teeming microcosm in a whole new and other world of life. Us humans, Joseph thought, with all our ongoing conflicts and complexity of mind are just a single species amongst the hundreds of thousands of extraordinary types of life.

*

The letter from Buck was alluring. Joseph mulled over Buck's words. Then he contemplated that tune *'Born under a wandering star'* again. Damn, he cursed silently to himself, why is Buck so often convincing? As passionate as he is in generalizing about his people, about their terrible plight and poverty in South Africa, his feelings do have years of experience. Yes, and his feelings are always sincere. It was also the intriguing way Buck had described the need for liberation in this apartheid country which had included Joseph's need to find his own freedom. It felt really alluring. In the way Buck compared Joseph's predicament to the underlying apartheid politics in South Africa. Joseph increasingly felt himself drawn to these parallels.

'I was born under a wandering star'. The phrase seemed to be taking root in a deeper part of himself. Echoing within Joseph's conscience, as persuasive as it was, it had begun to reverberate in his brain, speaking also of the enticement of Charles Darwin. This song-phrase from the Hollywood movie was even becoming a likely path to his own freedom: directing him perhaps in how to express his own talents. Perhaps like Buck he may

end up more alone than he had ever imagined. But it need not necessarily be the case of giving up friends and loved ones, he thought. After all his hero Darwin had enjoyed a family life.

Then from his growing apartheid conscience, like the anger of a twisting stomach gripe in his gut, he imagined all those other voices on the bus vying for themselves: the *non-white* majority in South Africa. Still *'non'* this and *'non'* that, still derided and mocked as citizens, still lined up and categorized by the authority of a racist government. Labelled the *'non-Europeans',* fit only to travel in the back of a bus or to live a stinking shanty existence. People forced to carry a pass in their own country and who were limited to zoned areas. They were the unshoed, unclean, unseen in this country. Unseen with a genius for tolerance.

Blind dependency: Buck called it. The privileged people in this country were both reliant and had created a reliance. And in the process they had blinkered themselves, no longer realizing the extent of their dependency and its inhumanity. They had cornered themselves in it. Becoming intolerant without their luxuries, oppressive and self-important with them, that Joseph felt he could agree with Buck that so much which has passed in South African history is based on this false sense of pride. This white supremacy as a force of dependency. This easy, loaded life of *white* privilege in a country which demands the compliance of *black* servants as slave labourers and gold diggers.

Joseph knew this personally because his family were no less reliant on low paid maids, gardeners, farm workers. The words in Buck's letter were even beginning to turn the tables on Mr Salem. By his own stance Buck seemed to imply that Joseph's father was just as reliant and dependent on binding Joseph to him, securing Joseph to the farm in his father's own name and service. A father who needed to be held to account, Joseph knew, a father who needed to accept the freedom his son was about to choose over and above the man's head. How would his father react? Indeed, how would he accept Joseph rejecting the inheritance of the farm? No one understood Harry Salem's motives better, and his wily ways, than his brother and ex-partner in the farm: Uncle Jack.

*

Again the intervention of Joseph's stomach, although this time the growing pangs of hunger for food and the grumble of digestive juices being secreted, forced him to postpone thoughts of Buck and of politics and his father, for later. He hoped for lunch at the homestead. Extracting his limbs from beneath the carpet of willow leaves on the riverbank, Joseph dusted off and rebooted his feet and emerged, blinking, adjusting his vision from the riverine shadows out into the sun.

The sweep of the Klein Drakenstein mountains towered around in the distance, making Joseph feel tiny in this place. Small yet familiar in these surroundings and quite at ease. The homestead at this vantage point was obscured behind a row of trees, facing east, perched at the foot of the Simonsberg range. The property of *Salemkop* farm lay within this upper Berg River valley: a wide, flat, expansive valley of winelands in the district of Paarl across which he was walking. Behind Joseph in the distance the bald granite dome of Paarl Rock was baking in the heat and exfoliating its outer stony layers like onion slices in the summer sun. The farm itself was of above average size: 195 hectares of vineyards with some additional plum, peach and pear orchards, and three plots of fallow land which were grazed in rotation. Crossing over a plot of fallow land, Joseph realized he had now traversed two sides of a triangle: *Kwaaiwater*, the university in Cape Town, the farm here at Simondium. The triangle would close when he returned to *Kwaaiwater* having finally made his decision.

He traversed a tract of land beside the river, land grazed recently yet not so heavily grazed as to be devoid of clumps of wild flowers growing in it. Twice his legs stumbled and sank down to the knees into mole-burrows. He avoided a third burrow hole in the soft ground, noticing beside the path a line of fresh molehills which belonged to the Cape mole-rat. In the distance was the familiar sound of giant refrigeration trucks passing by with a hollow droning sound as they transported farm produce to the town of Paarl. These heavy trucks, infrequent and noisy along the motorway, reminded Joseph of his boyhood and the start of a long-standing feud between his father and Uncle Jack. At a time when he was too young to understand the intricacy of this family feud. Beyond knowing it had something to do with that motorway road which split the farm.

For the three generations that had passed, *Salemkop* farm consisted of one main vineyard spanning sixty hectares, before the construction of that tarred road to Franschoek. Joseph was five years old at the time the vineyard was neatly sliced in half by a government road-building scheme. Although initially the family had reacted jointly with anger expressed both by Harry and Uncle Jack, no matter the government compensation paid, the road-building scheme began, ironically, an internal family feud which had never resolved. A feud increasing in complexity. A feud of internal Salem politics, inheritance, of farm succession. Today on his twenty first birthday as Joseph neared the tree-lined homestead, he felt it was his prerogative now to know why.

Rounding the vehicle yard and livestock paddocks he was met by a tail-wagging collie who recognized him instantly. The dog let out a yelp of excitement, circled Joseph once, sniffed him, then turned and headed off in the direction of the farm manager's office. Walking into a large, grassed courtyard enclosed on three sides by whitewashed outhouses, Joseph noticed the door was open to the farm manager's office. Then he saw Uncle Jack issuing instructions to a labourer. Joseph ambled along carrying his rucksack, the carboard box, and a heavy shoulder bag. The bag's strap weighed heavily on his shoulder as Uncle Jack signaled his approach with a wave of a hand.

"This *is* a lovely surprise," said the man. "I thought you were visiting *Salemkop* on Monday with our attorney." He reached forward giving Joseph a brisk handshake. "Happy birthday nephew!" Uncle Jack placed a hand on the shoulder of the labourer standing beside them, telling him in Afrikaans about Joseph's 21st birthday. The labourer acknowledged this information with a broad toothless smile, raising his flat-cap to Joseph and revealing a heavily calloused hand.

"Got a bit of time to talk?" Joseph asked.

Uncle Jack nodded. The farm labourer listened to Uncle Jack issue a series of further instructions in Afrikaans as Joseph stood by his Uncle, an unlikely looking farmer, tall yet slight-of-build and bespectacled. In his courteous manner and with an air of gentleness about him he dispensed a final directive. Once the labourer was on his way Uncle Jack offered to take the heavy bag from

Joseph's shoulder. Together they passed out of this semi-circle of outhouses *en route* to the farmstead.

"What are you carrying in here?" Uncle Jack tapped the shoulder bag. "It weighs a ton."

"Books."

The man shook his head. "Out here... on your birthday? You really can't leave it alone, can you!"

Joseph smiled. "They're the last of my things from university. I'm taking them back to *Kwaaiwater*."

"So you're finished with your studies. All set for the Monday signing then?"

Joseph hesitated briefly. "That's why I've come out here. I need to discuss it with you." Curling his thumbs under the straps of the rucksack at his chest, Joseph leaned slightly forward as he walked. "

Their eyes met in acknowledgement. Uncle Jack recognized the look in Joseph's eyes. And the hesitant tone in his voice. In recent years when training Joseph as assistant farm manager, they had developed a silent understanding. A shared family pact. An unspoken rapport. Certain Salem family questions about Joseph's father would not be probed. Historical animosities were best left behind. Out of bounds. Mr Salem's accident put paid to old, long standing criticisms. Uncle Jack showed kindness in this way. A restrained and quiet man himself, Uncle Jack was unselfish and unambitious and had often been misunderstood in the shadow of his pushy and extrovert brother. Now as they walked towards the homestead, Uncle Jack had the good sense to change the conversation, explaining that today his wife and two daughters were out visiting at a neighbouring farm.

"Fancy a bite, Joseph? I'm peckish."

"Famished. I was wondering when you'd ask."

He laughed. "If you don't mind leftovers from a guinea fowl meal we ate last night."

"Mmmm… sounds delicious. How can I refuse."

They approached the gabled house from the front.

"Fancy sitting outside in the garden under the pergola? You go round to it while I tell the maids we'll be eating out there."

Joseph caught his Uncle's arm before the man departed. "While you're in the house, can you find the farm contract and bring it out for me?"

Uncle Jack paused to consider the urgency in Joseph's voice.

"Sure," he said. "I expected you to ask for it."

*

In the whiteness of its white-washed walls and its traditional Cape-Dutch style architecture, *Salemkop* homestead gleamed in the afternoon sun, much as it had been a white beacon for over 250 years. The H-plan layout of the house was emphasized by its long parallel thatch roofs, and high sash windows each framed by shutters of olive green. Sitting in the back garden under the leafy pergola it was obvious to Joseph that anyone else looking in from a distance, visiting as a connoisseur of wine or a tourist exploring the wine route, would be honoured to inherit this unique place.

When Uncle Jack returned, the contract of inheritance turned out to be an aging scroll of parchment bearing the official heading 'Property Deed' inscribed in a curling old-fashioned script. Joseph examined the title which was brush-stroked in elegant letters on the yellowing parchment. Below the title, as Joseph unrolled the scroll, were paragraphs of legal jargon describing the farm and its original owner, its exact location in the valley, and its dimensions in outdated measurements. Following these and other legal paragraphs to do with later ownership and inheritance were flourishes of signatures made by descendants of the Salem family.

"That's where you'll sign," Uncle Jack pointed to a place with a dotted line on the lower part of the scroll.

Family signatures and the names of their witnesses continued down to where Joseph recognized his father's signature together with Uncle Jack's. An asterisk highlighted these last two signatures of the Salem brothers, which referred to a few lines of small print at the side of the page.

"Why the asterisk?"

Uncle Jack raised a hand in interruption, running a thin long forefinger across to a segment of text. "See these lines," he pursed his lips in emphasis.

Joseph read six lines of a quill-written script.

"That's the real crux of it."

ON REACHING ELIGIBILITY AT AGE 21, THE ELDEST SON STANDS TO INHERIT 'SALEMKOP' OR SHALL TAKE HIS PLACE IN THE LINE OF SUCCESSION. THEREON, HE WILL ASSUME DUTIES COMMENSURATE WITH BENEFICIAL UPKEEP AND ADMINISTRATION OF THE FARM.

"It's those few lines..." Uncle Jack reiterated. "They are really what you'll be signing up to, Joseph. In each generation the farm legally may be inherited by the eldest Salem son."

"You say *may*?"

"Yes, *may*. There is a choice."

"Oh..." Joseph went silent. Thoughtful. Then he pointed to the scroll: "What about the asterisk."

"You mean by our signatures? Pah, fancy you noticing that. There were certain complications around the issue that your father and I are twins. It confused the process of inheritance. Since there is no proof who of us is the elder twin. The small print marked by that asterisk on the document is a legal clause giving both your father and myself equal rights in *Salemkop*."

"So both of you signed happily?"

"We both signed up sharing a dream in the farm, yes. That was the theory. In practice, it became very different when we got down to working the farm."

Uncle Jack scrolled the parchment into two tight rolls meeting at the middle and tied them with a burgundy ribbon, placing it on a garden chair to make space on the table. Two maids had arrived from the farmhouse each carrying a tray of food.

"What a feast," said Joseph.

"Help yourself son."

"I vaguely remember..." Joseph said hesitantly. He watched the man slice slivers of game meat. Uncle Jack served a few slices of guinea fowl breast onto his nephew's plate. "Thanks. I vaguely remember when I was a boy, maybe twelve or thirteen years old, that a conflict flared up between you and Dad. Which now thinking about it had something to do with that road which bisects the farm."

"The tarred motorway?"

"I never understood it."

Uncle Jack sat back after placing two slices of guinea fowl onto his own plate. "A conflict? My God, tensions were high, Joseph." He spoke in a soft reasonable way. "Your father has such a competitive streak. Compulsively competitive. I experienced difficulties because his dreams for the farm were quite overbearing, but they had been there long before the road building scheme. The road, however, definitely focused them. The road gave him purpose. Added incentive. Tensions between me and him were apparent well before we ever inherited the farm. Being his brother, I know his personality well. With us being twins, growing up together, I always felt his need to dominate for as long as I can remember." Suddenly Uncle Jack raised his hand in opposition as if touching on one of those unmentionable subjects he and Joseph had learned to avoid. "Enough of that," he said in apology. "Why do you need to know about the road anyway?"

"Signing this inheritance is a serious issue for me. And I'm tired of seeing only half the picture."

"I'm not sure talking about it will help."

Joseph lifted a glass of red *Salemkop* pinotage wine and clinked it with Uncle Jack's.

"Cheers. You'd be surprised how much a young child picks up. You'd be surprised that I overheard you and Dad speaking once as a little boy. I remember after the road was built - just after - overhearing a conversation between yourselves in the wine cellar."

"Oh?"

"Dad's exclamation from across the cellar to you. I remember it exactly. Overhearing it while I was playing somewhere in a dark recess."

"You sure you want to tell me this, Joseph?"

"I think it has to be said now."

"Ok," Uncle Jack touched Joseph's hand reassuringly.

"Jack!" said Dad. You raised your eyes to him from a wine list you were busy ticking off. The smell in the cellar was musty, I recall, damp and the place made your voices echo. Dad walked up to you and said: "I was thinking, since it's time to prune the vines - you know I've always disliked the pinotage vineyard." He emerged from the shadows to stand under a light to face you. "Why don't we also plant hanepoort below the road now and so have hanepoort throughout, one type of grape only, in the west vineyard too."

"A-hh." Uncle Jack smiled. "Now I remember that conversation."

"That was the start wasn't it?"

"It was the germ of an idea that lead on and on, yes. Slowly, almost so slow as to be imperceptible, your father began to claim the property below the new tarred motorway. He wanted full authority over it. So illicitly he began to claim it. Secretively. Neglecting in the process many of the shared duties we had above the road. The orchards, of course, and the homestead were communal areas as they had always been, and they were above the road: for these there could be no such grappling against the concessions of the will."

"What about the pear orchards he planted below the road?"

"They also became a problem. One day..." Uncle Jack's expression turned pensive. He shook his head. "One day - you weren't there, Joseph – an argument ensued started by your father, about his need to occupy the homestead separately. He justified it by saying this was necessary since we were now partially working different parts of the farm, reaping some separate profits and your father believed he had all but secured the greater, more productive piece of land."

"How did you react?"

"I always believed in the equality of the inheritance. That we owned *Salemkop* jointly. Yet I always gave him the benefit of the doubt over it. He was so passionate, so unbending. Stupidly I realized the dangers when it was almost too late."

"Which was when?"

"After he drafted architectural plans to partition the homestead entirely, replacing original antique oak doors and doorways with brick walls. To try to create quite separate living quarters for our families."

"Goodness. I never realized that," said Joseph.

"And when ten families of labourers were forcibly moved by him to the old shanty village below the road, to work in what your father now called his 'new *Salemkop*' plum and peach orchards, I put my foot down. I telephoned our attorney. Which happened only a few weeks before that terrible fishing accident at *Kwaaiwater*.

Joseph drew his knife and fork together on the plate. He wiped his mouth with a serviette.

"Is Dad really that mercenary?"

"Whatever reasons he had for doing what he believed, your father was unapproachable, unbending to any discussion and powerful in his passion. Unreasonable in his claims. I'm not sure whether I could ever have stopped him single-handedly. It was the fishing accident and his paralysis that put an end to the dogged way he fought to create a separate claim in the farm. Which, from whatever angle you look at, is a claim against the inheritance."

Joseph's eyes were serious.

"You know," Joseph said, leaning an elbow on the table, "I doubt his days of trying to stake a separate claim are over."

"How do you mean?" asked Uncle Jack.

Joseph shrugged. "*You* know Dad as a brother. I know him as a father which is only mildly different. Subtly for years he has been grooming me to become his faithful lieutenant. I'm sure to him my birthday means that I will take up this farm battle on his behalf."

"But you have a choice, son."

"It may sound absurd to you but choosing otherwise will make me feel as though I'm robbing him."

"What do you mean? That you feel that much guilt?" From across the table the man looked questioningly through his spectacles.

"Uncle Jack, how would you feel if I continued my studies at university and decided not to sign the inheritance? If I did not become a farm manager with you?"

"Entirely your own choice. That is YOUR choice, Joseph. Anyway it's not as if there are queues and queues of Salem males lining up to take your place: at this point in time you're the only one eligible. What's the rush? I've still got at least a decade of farming in me."

"Hah, try tell that to Dad. What on earth do I say to him? He's raving and ranting about how he can't lose this inheritance. How I must be there for him, as a son, to protect his rights, to protect his farming property. He implicitly believes that he needs an ally here at *Salemkop* - that ally being me - to stop you from usurping his claim in the farm."

"Nonsense! That's typical Harry. That's my brother. For one, you're not robbing him of anything if you don't sign. This farm is part of the family."

"Then why do I feel so guilty?"

"Joseph, I can guess but I'd prefer not to. Suffice it to say that in many circumstances I also don't see eye to eye with your father. He and I have always tackled things very differently. Let me tell you: in the end all we really have in

237

life are our principles. And the actions that protect and extend and guide those principles. If we ever are judged, it is these principles by which we will inevitably be judged. Not that principles don't ever change. Of course, principles evolve. That is why we should always be striving for a greater wisdom, truth and humanity in those principles."

The young man sat back, astonished at the breadth of what he was hearing. A confession about his father. An expose uttered by his reticent, introverted uncle, who always, for as long as he knew, preferred to remain quiet. Silent. Unheard. Unseen. An uncle whose politeness, kindness, temperate nature, always seemed just skin-deep.

"Joseph," Uncle Jack continued: "this may sound *a bietjie* sentimental. But if there is such a thing as democracy of feelings, democracy of caring, then strive for that. You're my nephew. I don't want to see anyone ramrod you with their pity or their disabilities. Son, don't allow anyone to make light of your dreams."

- CHAPTER 10 –

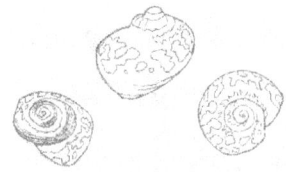

Despite Joseph's unexpected visit and sharing a lunchtime meal, Uncle Jack's time was limited. Farm hands needed him in the north sector of the hanepoort vineyard. *En route* to this work, Uncle Jack collected the keys of a small pick-up van from the farm manager's office and locked Joseph's rucksack, shoulder bag and cardboard box in the vehicle. Joseph strode with him in the cool lengthening shadows thrown by outbuildings and homestead trees to the perimeter of the vineyard, waved his Uncle off, and himself crossed to the *agterpad:* a rear path leading through the upper plum orchard to the foot of their *kopje. Salemkop* derived part of its name from the dome-like *kopje* in the south west corner of the farm. From the crest of this hill the intricately laid out vineyards and rolling Cape beauty of *Salemkop* could be viewed out to its boundaries and beyond to the larger valley and to other neighbouring wine farms.

The orchard, as Joseph turned into it, gently sloped up the valley side. Rows and rows of plum trees stretched out straight ahead of him in near perfect v-perspective lines. The *kopje* was like a primeval monolith burst up from the earth and it wore the only coat of wild indigenous vegetation around these cultivated fields, all the way to the Simonsberg range of mountains. The plum trees would soon be heavy with ripening fruit and the branches sagging under their weight. Joseph stopped in his slow climb up the slope. Crouching down in the shade of a plum tree to rest on his haunches as a small flock of guinea fowl approached at ground level. They passed nearby chattering at each other, scratching at the dry sand with their feet, creating a small dust cloud in their wake. Grey in colour with white speckles the attractive plumage of these wild fowl contrasted with the small head which carried red fleshy wattles. Joseph counted eight adult birds with five chicks in trail. He resumed walking, his vision lowered to carefully watch for adders or scorpions.

Being on the *agterpad* and heading towards the *kopje* felt to Joseph like a birthday treat. How peaceful it was. How strange it seemed. Because all the questions and dilemmas considered over the past few days ended right here at the farm. Lexa, Buck, Professor Harris, Uncle Jack, had now offered their advice. Each in their own way had, sincerely, helped to unwrap a world containing fewer conditions and family rules which was hitherto an alien experience for Joseph. These people who seemed to care offered a brighter alternative to the regime which his father upheld at *Kwaaiwater*. In revealing its face, another option, another world, had begun to speak to him with a new and softer kind of reassurance, above the suspicions, like a hand proffered holding an olive branch.

Squinting into the sun as he reached the orchard's southern edge, Joseph turned right, following the last line of plum trees in their flow along the perimeter up to the *kopje*. He was perspiring lightly into his shirt, his brown cotton trousers pulling tight at every stride and a prickle of sweat itched at his scalp. The *kopje* marked a clear outline this mid-afternoon, rounded and symmetrical, but Simonsberg towering behind it was silver in the distance and unclear in its craggy mountain profile as it loomed precipitously in the heat haze.

At the last plum tree Joseph's step kicked a stone and a locust suddenly leapt from a grass tussock. He watched its noisy whirr of wings taking the insect high and safe, the sound a 'click-click' buzz, and it made him think of the poster in his bedroom of the leopard-baboon encounter: that prodigal baboon facing a life-or-death ordeal head on. Then the locust landed and it was in the wild indigenous *fynbos* and Joseph had reached the foot of the *kopje*. Although still a schoolboy in those days, it seemed just months ago since he had measured the study transect up this hillock: a band of nature which he'd demarcated for observation and experimentation that stretched from the base to the summit and for years had served as his scientific playground.

Joseph threaded his way up among the boulders and the fine-leafed bushes. He remembered choosing a special notebook in which to record this childhood project, dividing the pages into section headings: geology, indigenous plants (family, genus, species), alien invader plants, soil-types and humus content, resident insects, scorpions, spiders, resident reptiles, birds, small mammals. He set himself the task of collecting specimens to identify the flora and fauna in this transect after weeks of reading up the proper scientific methods used. Scientific texts taught him to mark out smaller squares to make a finer study grid inside the main one.

It was fun to sink plastic buckets into the ground as pitfall traps to collect and identify surface animals moving about, before releasing them again. Bushes were shaken for insects which dropped into nets positioned underneath. A bird-watching schedule was set up in a purpose-built observation hide. Rodents and shrews were live-trapped, toe-marked and then set free, to study their movement across the study area and home-ranges. Stones were delicately upturned. Burrows investigated. Termite mounds counted and measured. Even the subject of palaeontology became necessary for the project when, halfway up the *kopje* and one year into his study, three stone-age hand axes made by prehistoric humans were discovered.

Finally, beyond this reverie of past childhood times, he reached the crest. Joseph chose a route which purposely did not follow the line of his old transect, though up on the flattened summit there were still a few markers of the old sample sites easily recognizable. He remembered having scrutinized this stretch of ground so intensively, with such studious childhood determination, that even the shapes of rocks and patterns of lichen markings were familiar to him still. Some old stone beacons remained standing upright in places along the transect perimeter, which served as boundary markers made of rounded pebbles stacked one on top of the other.

Joseph sat down on a flat rock, beads of sweat on his upper lip. *Salemkop* homestead from this vantage point had dwindled to the size of a toy glistening white, affluent, H-shaped, with the porch visible and pergola where they had lunch overlooking the back lawn and garden. From here he could make out the window of his old bedroom. But the homestead was secondary to this *kopje*. This *kopje* had spurred Joseph's reality as a child. It had been his playground, his boyhood castle, the earthy subject of his dreams, over and above any other place or thing. During those childhood years, he learned what made this *kopje* tick: in its geology and its natural history. And by spending the greater time here, Joseph felt he himself was reflected in the life of the *kopje*. His personality had germinated on this African ground, urged on by sunshine and crickets, birdsong and the oily scent of wild bushes. Those cultivated fields below never promised as much as this wild hill, nor did the Cape-Dutch style homestead.

A small figure of a man could be seen walking about down in the hanepoort vineyard. Joseph shaded his eyes against the sun, screwing them up for better focus. It was the figure of Uncle Jack pointing at a puff of white dust. The dust cleared. Then another puff of white. A second figure appeared out of this dust with a bag in hand. Scanning the rows of vines Joseph could now see three more dust pockets rising in different places. It was sulphur dust, he guessed, caught in the breeze, instigated by three more figures of farm labourers. It was impossible to recognize their identity. He was too far away. They must be busy filling in on yesterday's sulphur dusting, Joseph thought, as the vines had paled under their efforts. The heat of the sun would soon transform this chemical dust into a protective fungicide.

Looking beyond the hanepoort vineyard lay that contentious tarred motorway. A string of ant-sized people walked beside it: could Johnnie be out there heading them, he wondered? Joseph smiled at the vision. If anyone was leading them, it would most likely be Johnnie. Religious Johnnie. Conciliatory Johnnie. Johnnie with the filthy face, darkest skin,   brightest smile, the orphan teenager in frayed shorts who lived in the labourers compound.

"*Baas,*" Johnnie would say. "Please *baas,* may I have a quick word with you?" It took less than a nod to prompt him - to set him off – circling you, breathing into your ear, trembling with the fervour of religious anticipation. His animated display in Afrikaans dialect included the chanting, rhythmic, glorification of his mighty God with whom you shouldn't play games:

> *"Die Here' is magtig, die Here' is krag*
> *Dis waar - hy is groot!*
> *Hallelujah Lord; dankie, baie dankie Lord*
> *God is die Lord, almagtig God:*
> *Jy moenie met hom speel nie!"*

Kneeling in front of you, wide-eyed, animated, Johnnie would enquire with a broad smile: "Can I pray for you, *Baas*? No matter whether you are the master or I the servant, I am in your service Baas for the coming of our Lord."

That was Johnnie, innocent and indomitable knight in shining armour, a barefoot orphan whose determination to look on the bright side Joseph had always admired. Who so clearly, so early on, had chosen his own pathway in life. At this vantage point high up on the *kopje* Joseph watched the ant-sized row of people walking away along the motorway. And it again made him think of that prodigal baboon in the poster on his bedroom wall and how Johnnie too had taken up the cause to stand and fight the total onslaught of the leopard.

*

Beyond the motorway lay a question. The question Joseph considered in the bright hazy sunshine over Simonsberg was thus. Why was this spelling out by Professor Harris and Uncle Jack of what Joseph should consider and what should be done, why was it coming from men who seemed to have fallen short of living by the standards they themselves espoused?

Don't swerve from your dreams each had urged, don't swerve from your principles and talents. Yet Professor Harris admitted that he had himself deflected away from his own search for scientific discovery. While Uncle Jack, once a keen land surveyor and on the cusp of starting his own business, had accepted his lot of inheriting the farm. And then as farm manager was swayed perilously by his brother.

Even Lexa this morning revealed the influence of her own parents demands to make light of her drama career and be more earnest about marriage. The only people he had spoken to who were actively living the conviction of their *own* philosophy, who stood upright in the knowledge of their *own* free choice: were Simon Thandiwe wall-papering his shanty with the flame of apartheid flickering in his eyes; and Buck wandering alone fishing the great Southern Ocean.

Joseph also felt puzzled by Lexa's attitude this morning before they freed the sunbird. The thought of it made him inwardly furious. When she suggested, up on the West Lookout, that he at least consider selling out to his father's constraints. Perhaps her emotions were stirred because their relationship had become intimate, he wondered? This morning there seemed a new vulnerability to Lexa's words. But if she and Uncle Jack could slide away from their own advice as Professor Harris had, capitulating in the face of demands, discarding talents or principles to the dust, then what did they have left to fight for? Other people's dreams? Perhaps it's a commendable attribute for other people to have, Joseph thought? It did not, however, resolve the driving force of his own talent. To fight for the truth of his own dreams, over and above dreams that others were giving up, or had given up, themselves.

Beyond the motorway the spurned pinotage vineyard which his father had attempted to destroy and replace with hanepoort grapes only, but had never quite accomplished, was still a patchwork of disarray. Remnants of Harry Salem's intolerance and persistence: a scar of earth, a weal from the family feud, a blot on the farm landscape. This damage at *Salemkop* had been cleared and the replanting of this scar was now reaching completion with new pinotage vines re-laid over the original site. The old shanty village nearby was once again vacated, the ten families of farm labourers briefly forced to occupy it by his father had permanently deserted the place for the main staff quarters.

So not only, Joseph thought, had this destruction of dreams happened to the *black* majority in this country, and now almost to him. But there was Lexa, Uncle Jack, Professor Harris who at least had felt the odds against them: this impact of human choice or lack of choice. The abuse of talents, the interrogators white face, the uprisings against apartheid around the country happening now, the fettered albatross, the leopard's green eyes, South Africa as it was at this time...

Joseph mused pensively on this, alone, as he traveled in the borrowed pickup van with his bags beside him all the way back to *Kwaaiwater*.

*

Upstairs in his bedroom at *Rus en Vrede* the view out across Walker Bay remained unchanged, mist-soaked and calming, but Jessica's state of mind had transformed. She trod into Joseph's room as a mane of black hair, as dark as the space around her, hanging loose, tumbled, dejected. You could always tell when she was overwrought by the way she hid behind this mane of blackness. And by the way she ran her fingers in repeated strokes through the locks, nervously, yet unconscious of it.

"Jessica, what's wrong?" Joseph asked, seated on the bed.

"He's locked away my music."

"Who?"

"Dad!" She melted down into a chair.

Joseph got up, his stomach threatening to coil into a spasm as it had threatened on so many occasions over these past days whenever negative thoughts of his father emerged. He took a few deep breaths to counteract it, his eyes shifting for an answer. When he reached Jessica, she leaned her head against his torso as he stroked the black mane, feeling the tenseness and dejection in her frame.

"What brought this on?" Joseph was surprised that when his words emerged, they sounded snarled. In response to her brother she composed herself and suddenly relaxed back into the chair, crossing one leg over the other. Pushing her hair away from her eyes, she smiled a half-smile as if embarrassed, then glanced down at the floor. Joseph considered this scene thinking to himself: God, we Salem's are so damn expert at re-establishing our composure, blotting out our feelings, while our insides wreak havoc.

"I'm not sure why he took it." Jessica controlled the quaver in her voice.

"What happened after I left this morning?"

She fidgeted with bangles on her wrist which made a tinkling sound. "I heard them arguing: Mother and Dad", she said. "Dad yelled something about *Salemkop* and you know how Mother tries to calm him. I couldn't make out why he was so upset, not above the sonata I was playing." She sighed and shrugged despondently. "Now of all things he's confiscated my music."

Joseph placed a hand on her shoulder.

"I can't believe this. It's ridiculous. You have my word Jess, I promise we'll sort this out. You say he mentioned the farm. Then what?"

"I was sitting at the piano in the lounge. Dad sped in from the porch, I couldn't believe it, driving that wheelchair like it was an off-road scrambler. 'How many blasted times...' he shouted at me, snatching the book of sheet-music, then disappeared into his bedroom. He hasn't come out yet."

Joseph glared venomously at her.

"And Mother? What did she do?"

"She saw my face - upset, crying - I think. She looked so trapped and confused..."

"Damn. Damn!" Joseph cursed.

"She told me that she would try to recover my music."

"Well has she?"

"Would I be here if she had?" Jessica defended herself, jutting her chin and putting a clenched hand on her waist. Joseph unfolded his arms to assume a less threatening attitude over her.

"Jess I'm sorry," he appealed in a low voice, tilting his head, realizing it wasn't her fault. "You say he locked your music away?"

"The argument started up again when Mother went into their bedroom to retrieve the music book. Dad shouted 'HE would decide when to return it.' Later on, apologetic and rather embarrassed, Mother told me that it's been locked away in the yellow-wood cabinet."

"So you haven't asked him yourself?"

"I'm scared, Joe." They glanced at each other knowingly. Joseph walked to the bed and slumped down.

"Why did you get dragged into this!"

"For playing the piano?" Jessica frowned.

"Look Jess, if I must criticize, and I don't like to, by now both of us know that Dad sometimes has the most irrational and intolerant reasons for doing things." She bowed her head as Joseph continued: "It's totally unwarranted for him to involve you because I'm sure the farm inheritance is behind his temper tantrum. You should never be mixed up in it."

"You mean he's still on at you about signing those papers from your discussion at breakfast yesterday! But I thought you sorted that out on the verandah?"

"I have until tonight to finalize my answer."

"Oh, ok..." she said surprised. "In the meantime, how do we recover my music?"

"I hate the situation you're in," Joseph said, his voice cracking with agitation. He glanced up at the leopard-baboon poster above his bed, then back at Jessica. "Has he ever done this to you before?"

"No," she mumbled into her hair.

Joseph slowly shook his head, a glimmer of sarcasm in his eyes.

"Must be another strategy he's invented, pulled out from under his wheelchair blanket," Joseph quipped.

Hearing these words Jessica glanced at her brother confused, finding this criticism harsh, her youthful innocence still effectively blocking out the impact of these ugly family dramas happening around her. Issues she preferred to avoid. She shrugged at Joseph defensively.

"Jess listen to me," Joseph pleaded. "I have no desire to build animosities against Dad or gang up on parents. Each of us needs to learn openness. To be more honest with one another. Understand each other without taking sides or being so involved in our own little worlds to the exclusion of all else. But now you're involved. And I feel it's my duty as your older brother to protect you. Explain it to you."

Her widened eyes looked innocent.

"Since Dad's accident we've all had to cope one way or another with his predicament. Without us ever really discussing to each other how we feel, without us talking about how it weighs heavily on every one of us or affects our lives. How we feel about Dad demanding things he does. And how his unreasonable demands affect us: if we feel ridiculed or overruled by them. Perhaps the time has come for you and me to open our eyes to the way he manipulates us."

"He's a sick man, Joe. He's got heart problems!"

Joseph nodded seriously. "You don't have to remind me. But that doesn't alter the fact he uses his illness - whether it's his weak heart or his paralysis – to gain pity and sympathy. To make us feel eternally sorry for him. Guilty even. Beholden to him. Manipulating us sometimes to do unreasonable things. I don't see much democracy of choice or democracy of feelings in this family. We're all adults yet most of the decisions are made unilaterally by him."

"What do you mean?"

"Just think how we all react around him when he gets flustered or demanding. Mother's scared, you're frightened, we all live in a perpetual state of fear. The situation is like a fascist state!"

"Don't say that."

"I MUST! I MUST point this out to help you understand another point of view." But this fascist state analogy had the effect of blocking her off, drawing her away from the conversation. As if this implication was too

253

unpalatable to digest. Nonetheless Joseph persisted: "Jess, the entire farm issue between Dad and myself rests on a principle," he said, emphasizing his point with a raised index finger. "Just one principle. Whether or not to pursue my talents."

"What has that got to do with me?"

"It has everything to do with him taking away your music."

Her eyebrow raised with renewed interest.

"Dad forcing me to take up the inheritance and become farm manager without my consent is not much different to him snatching your music book. He is manipulating me from making my own choices to discover what I truly want for my future."

"That's not what I believe Dad was doing." Her expression turned to one of astonishment. "He really does care about us."

"Jess, please promise me that you'll never allow him to persuade you against your music career."

"Why would he do that? He understands what music means to me."

"He's already testing you by taking away the sheet-music."

"Maybe I was practicing too much. Or too loudly. Maybe I was getting on his nerves. I don't know, he just snatched it away."

Joseph held his hands open to her in appeasement. "Alright. Later on, when he and I discuss the farm, I'll ask him for the music. But how about us making a pact: you and me, Jess," he said, looking into the pools of her dark brown eyes. "To always respect how important our talents are to us. That you will protect your music career, hold onto your dreams, and always remember the principle of it."

Shrugging her shoulders without clear conviction, apparently unconvinced by this threat of her father, Jessica stood up and went downstairs.

*

So, in the end, Joseph contemplated inwardly while walking around his room, we are all alone with our decisions. As he paced, he pushed aside scientific implements hanging from the rafters. Ultimately, we are the ones in the end who must live with our decisions. He relaxed back onto the bed, his thoughts now cruising and quite free of regret. It was a peacefulness of mind which made him recall words his father had once read to him from his hospital bed. The words of proverb 9.12:

> 'If you are wise, you are wise for yourself;
> if you scoff, you alone will bear it.'

Hah, he wryly shook his head. How would Dad react to these very words if he – Joseph himself - used them, flipped them around, to counter his father. Used them as motivation to reject the farm inheritance? Anyway did it really matter what words of motivation he used? Making an excuse seemed like he was being defensive. Why did he need to excuse himself? After all this career decision, once confirmed, would have to be lived, experienced, not just signed and written down on paper in some kind of vacuum. Young and inexperienced as he was, literally only starting out, this step would determine his path for years to come: there was a freedom and sense of power in having this choice. Yet even if he blotted out the parental pressure, he still had a sense of hesitancy. It was hard. Joseph realized that no decision is merely *black* or *white*?

It seemed paradoxical his father's attitude was teaching him more about this apartheid country – about *black* or *white* - when his father perceived so little of it. Perhaps that was the problem. As Buck put it: blind dependency. Allowing yourself to believe this exclusivity, *white* or *black,* surely makes you a poor decision-maker? If you can only decide by simplifying life into one or other point of view. Distorting the complexity of the whole truth in order to try to understand it! That is how South African society had constructed its human barriers. Joseph had also reached the conclusion that neither was having a talent absolute. It is often a fragile thing in need of being protected. But if nature gives us a calling, no person or society has a right to deny it.

There were other considerations too which Joseph knew could just as easily justify him becoming farm manager. Over and above the outdoors rural lifestyle which in itself was an attractive reminder of his childhood: the open spaces of farmland, wilderness of the *kopje*, the Simonsberg range towering in the distance. It was precisely these things which gave him a taste for biology in the first place. Observing ibises glide by in noisy flocks over the farm reservoirs; the love-chases of turtle doves; the cattle egrets stalking ticks in the long grass; the glowing dusk reflecting its shades of orange and pink onto the Klein Drakenstein mountain range. In a large part he owed this appreciation of nature to his father who had brought the family to *Salemkop* in the first place. As it was also his father who funded his university education and who trained Joseph on the farm with such expectation.

But his considerations of inheriting the farm now went further and deeper than simply inheriting a wine farm, in light of his growing apartheid conscience. Managing *Salemkop* for Joseph no longer merely meant simply harvesting grapes, fermenting the product, bottling the wine and selling it. Now he knew how much added authority and influence he would have over the *coloured* farm labourers. It was an authority that could enable him to start projects to educate and help the children of Johnnie K, Marietjie, Hannes, in fact all of the farm workers. He would have the authority to raise their standard of living out from a level of poverty.

And especially, Joseph thought, I could speed up the programme of rehabilitation which has been so desperately slow in bringing our labourers out of the alcoholic habit of the *tot system*. The habit of alcoholic enslavement to which farmers subscribe. From the bondage of the *tot system* that for so many generations has paid labourers in tots of wine for their work. A traditional method of payment which created generations of subservient farm workers totally reliant on alcohol. Uneducated. Inebriated most of the time. Working in a haze of alcoholic stupour. Wine for breakfast, lunch, supper. Inebriated by their trade, making these *black* and *coloured* alcoholic workers entirely dependent on *white* farm owners. A convenience for farm bosses who, un-thinking, served well by this system, perpetuate the cycle of alcoholism from labourer father to labourer son.

Joseph even imagined how becoming farm manager may give him a chance to smooth over soured family relationships: perhaps even get Uncle Jack and his father to shake hands.

On the farm, Joseph mused to himself in the quiet of his laboratory bedroom, think of what a power of good I could do.

- CHAPTER 11 –

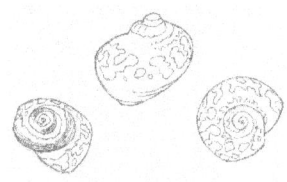

With his palms resting on the sill of a sea window, Joseph looked out from his loft room at *Kwaaiwater* to the distant horizon. It felt as if this closing sunset, still an hour and a half away, was signaling the birth of his career. Certainly, by sun-up tomorrow after a weekend of urgent soul searching, the decision would be a *fait accompli* and half a day old. A flock of cormorants finished with their fishing for the day were flying home low over the water in an undulating v-formation. They resembled a flight squadron returning from battle. Joseph didn't even try to understand why his blood felt warmed by a sense of achievement; his whole body felt bathed in a strange calm, his thoughts now as tranquil as the panorama across the bay.

Even though he knew the time was near to approach his father, this unusual calm lingered as Joseph descended the stairs. A calm rose from his legs as a kind of numbness exploding upwards, a mushroom cloud tingling the base of his spine and turning all his thoughts inward. He felt like a totally self-contained being descending the stairs; like an Orpheus in an hypnotic trance. Or an island risen from choppy waves to do unknown battle with the wind and the sun. He left his bedroom without any last minute preparations – his thoughts silently spent - his mind geared into neutral as if telling his body to 'TAKE me where you will, DO what is intended, teach me how TO BELIEVE'. Like an exhausted athlete having just run a cross-country marathon it felt as if he had been told to turn around and run back, his muscles and mind singing the limit of their endurance. This weekend indeed had been a mind race. A mental exhaustion. And now there was this unusual feeling of acceptance.

Unexpectedly his mother exited the lounge carrying a watering can. "Joseph, you're back!" she exclaimed. Tenderly she placed an arm around his waist. "Come with me into the kitchen and tell me about your day." Her touch had the effect of drawing Joseph out of his almost sleepwalking state. They passed by the main bedroom where Mr Salem would be resting and Joseph noticed the door closed, unusual even for a Saturday.

"Cup of tea?" she offered.

"Thanks." Joseph nodded. From the kitchen he glanced down the passageway at the closed bedroom door again. "Is Dad asleep?"

"He's lying down."

"I need to speak to him."

"First have some tea," she said with a smile and a twinkle in her eyes. "I've baked you a surprise!" Her enthusiasm was almost child-like as she pointed to an object hidden under an embroidered white doily. She carried it on a tray to the kitchen table together with a pot of blackcurrant tea. Sitting down opposite Joseph, she poured the tea into two china cups. "I hoped for some time alone with you before your party. We haven't seen you the whole day. How did the packing-up go at university?"

"Quite well actually. It didn't take very long. After three years I'll miss that residence room."

She half-smiled. Her expressive liquid grey eyes always conveyed a hint of the emotional, they always seemed undecided in their mood as though midway somewhere between tears and joy and a gentle reconciliation.

"Alright now!" she said with a dramatic turn of voice as her hands went to the doily. "Are we ready? Close your eyes." Joseph closed his eyes to add more tension to the unveiling. "There..." she lifted the doily, beaming with delight as he opened his eyes, lowering them from her face to fix onto a small iced cake in the shape of a volcano. On its surface was a heavy sprinkling of nuts and crushed mint crisp.

"What's this?"

"I couldn't resist. It's our *kopje* on the farm," she explained. Turning it around, the name *Salemkop* written cursively in whipped cream. She watched Joseph examine it more carefully with a degree of astonishment in his expression.

"Would you believe it," he confessed, "I've just come back from visiting the farm. And when I was there I climbed the *kopje*!"

"Well quite frankly I must say that before baking it I racked my brain thinking of what would stand out most in your childhood memories that could possibly symbolize you becoming an adult? And for me it couldn't be anything other than our *kopje*."

"May I cut it."

Smiling, Mrs Salem quickly, enthusiastically, handed Joseph a knife. The surface of nuts resembled boulders and rocks on the *kopje* slopes. There was even a sprinkling of crushed mint crisp representing the *fynbos* vegetation. It cut smoothly, easily, and Joseph served his mother the first slice. Before cutting the second slice for himself he hesitated, feeling distracted, staring down hypnotically at the cake because it was uncanny. The rectangular gap from summit to base he had just removed in the first slice were exact dimensions of his old study transect, perfectly to scale, even significant in the position of the cut. The whipped cream letters which spelled *Salemkop* remained untouched facing his mother, whilst the image of the transect hovered up at him like a question mark.

Whilst his mother was oblivious of the significance to Joseph, he remained transfixed for some moments as she waited for him to cut a second slice. Some minutes passed. "Joe," she said finally, "I'm not starting without you!" He snapped out of this reverie. He took a sip of blackcurrant tea. Eating the cake, they began to discuss the farm: how it looked at this time of year and how was the family. She asked fondly after Uncle Jack; her curiosity even stretched to a mention of orphan Johnnie. However, in this discussion, Mrs Salem's liquid grey eyes seemed to harbour something heavy weighing on her mind.

"Jessica told me what happened this morning," Joseph confessed. "To her music," he added. He guessed this as the cause of her weighted expression. Immediately she cast down her vision.

"A ghastly time we've had of it," she replied. "Your father has been jittery this whole weekend after your conversation with him about the farm. Joe, it's been enormously difficult for me. Jessica seems to have become a scapegoat. It's the first time he's ever locked himself away in our bedroom. He's been there for half the day now. He won't let me in." Her expressive eyes filled with tears, her lower lip developing a gentle quiver, yet she remained gracefully composed in her sadness. She took a last sip of tea. She laid the cup down on the table.

"Do you think he will let me in?" Joseph asked.

"It's certainly worth a try. At this stage you are the only one who can defuse this issue."

There was a shared silence between them.

"Be gentle with him Joe," Mrs Salem broke in.

"Thank you, Mother, for the most unexpected cake and all the trouble you've taken. I'm sorry it was so difficult, so stressful for you today." He circled the kitchen table and took her frail shape in his arms. When holding her against him, he could feel her body was so tense with emotion that she was trembling.

*

At the closed door to the master bedroom, Joseph paused, his senses strangely becoming enveloped again in a mushroom cloud of numbness. This calm was indeed strange. Calm enough to arouse his own inward suspicions. He contemplated the hackles raised along the spine of that prodigal baboon in his bedroom poster. At that point of head-on conflict there seemed to be so much action, movement, terror, so contrary to his present calmness. He waited at the door. When thoughts of his tearful mother returned, Joseph suddenly realized that she was quite like him. He couldn't pinpoint exactly where the similarities lay, or why he thought it, although in her eyes... he didn't want eyes like that. In their fluid grey beauty there was too much defeat.

Joseph knocked briskly. No answer. His vision fixed on the doorknob, expectant, wary. Then from his body numbness came a faint, almost imperceptible, stomach twinge. The doorknob and anything inside the room still did not move or make a sound. He knocked three more times.

"Dad, it's me Joseph," he called out.

A faint shuffling noise emanated from behind the door.

"Who's that?" The voice sounded tired.

"It's me. Joseph."

There was another bout of shuffling.

"Come in."

Joseph turned the doorknob. The room was dim from a drawn curtain, yet there was enough illumination from the strangled light to distinguish faces in framed photographs on the dressing table: Jessica as a young baby, a teenaged Joseph riding a horse, the family together in their youth, a breathtaking view of the farm taken from high up on the Simonsberg. Mr Salem looked half-awake lying down covered in a patchwork quilt, his wheelchair beside the bed. The man had obviously just pulled himself up into a sitting position on the bed because the cushions behind him were not arranged comfortably and his paralyzed leg was uncovered and pointing at a grotesque angle.

"Dad, can I make you more comfortable?"

"The cushions," he wheezed. "Puff them out for me." Mr Salem lifted his torso on his powerful arms, then leaned back onto the rearranged cushions, sighing and quite oblivious of that crumpled limb. Tactful not to cause embarrassment by offering to help with the leg, Joseph simply pulled and patted the quilt straight and covered it up.

"Take a seat," said his father, eyes still blinking and encrusted with sleep. He pointed to the wheelchair as a place to sit, and Joseph climbed into it, not something he often did. That done, they were oddly side-by-side facing the door. It was not unlike the side-by-side communication they shared when meeting on the porch during their chats looking out onto Walker Bay. "Son," he said softly, turning his head in the confined remit of the cushion. "I haven't yet wished you happy birthday." He reached for Joseph's hand on the wheelchair armrest and held it tight. "My blessings to you."

"Thanks Dad."

Joseph felt the powerful squeeze of his father's hand.

"I went to the farm today, Dad."

The man's heavy shoulder twitched at these words and his handhold went rigid.

"The orchards are looking good for the peach and plum harvest. Looks like a better crop this year. Uncle Jack was in the northern sector of the hanepoort vineyard checking how well the sulphuring went yesterday. I had time to climb the *kopje*."

"It's looking alright is it?" His grip on Joseph's hand eased. "Did you manage to talk to Jack?"

"Briefly over lunch. He was occupied with work."

"Did the vet examine that cow?"

"I didn't ask."

"Why not?" Mr Salem raised his voice.

"I forgot, I didn't think of it. I didn't visit the farm to ask about that cow." The grip on Joseph's hand tightened and he dissolved into silence. Mr Salem had a short bout of coughing, then he too fell silent and they both sat staring blankly at the closed bedroom door.

"Harry, are you awake?" Mrs Salem's worried voice emerged muffled from behind the door.

"I'm busy talking to Joseph. Please don't disturb us," he said. Under the door the two men could see the shadow of her feet. Shortly, the shadow disappeared with the patter of footsteps.

"Son, I don't appreciate your sarcasm."

Joseph did not reply. Neither did he offer a sideways glance towards his father.

"It hasn't been an easy weekend for me you know. I've spent half of today locked in this room sleeping and worrying about you."

Joseph turned to him. "Believe me if I say it's been just as difficult for me?" The iron grip on his hand was becoming painful. "Dad, please let go of my hand." His father's hold lightened but remained firm.

"Did you hear what I just said?" Mr Salem's voice was raised, with an edge to it. "That the whole day I've been worrying!" Another silence, pregnant with waiting. As if the man was waiting for a sympathetic reaction. There was expectancy in the way he was sitting.

"Dad, have you ever climbed the *kopje*?"

"What!" Mr Salem's voice was incredulous.

"Today I climbed the *kopje* and wondered if..."

"You wondered, did you really! *Kopje, kopje.* I'm sick and tired of you mentioning that *KOPJE.* Can't we have a decent discussion about the farm without it becoming irrelevant?"

"It was a birthday treat to myself today to climb the *kopje*. It has special memories. Of when I was younger, of my childhood, and the place still has enormous significance for me, even now."

"It sounds a tad childish to me. Where is the man I raised, Joe? The twenty one year old you've become today?" With that Mr Salem's grip tightened over Joseph's hand until the knuckles cracked and he let out a muffled groan. His father would still not let go until Joseph painfully wrenched his hand free.

"Enough!" Joseph's anger was up. "Alright!   You're right, I am twenty one. No longer a child. So, stop treating me as though I am unable to make a decision."

"Well if there's a man behind that childish face of yours, why isn't your mind made up by now?" Mr Salem's eyes glinted with menace. Joseph paused at the fire in his father's eyes, frown lines on his forehead, clenched jaw. Then he answered quietly, without being baited, holding out to this menace and without malevolence, he made eye contact with his father and in a polite tone remarked:

"My mind is clear. My decision is made."

*

The art of interrogation, in South Africa as elsewhere, is seldom a subtle art. Indeed, when the boundaries of subtlety blur into the coercive, it is perhaps questionable whether the interrogator is an artist at all. In the courtroom there certainly are artists, and they abound on the political stage, but locked away from the public eye - just victim and interrogator - this art, if indeed it ever is art, has no need of civilization's trappings. Nor the moral niceties of public politics. Nor the tolerance of outward public appearances.

Therefore, closed off to outside interference and shielded in his own bedroom, Mr Salem could easily vent his disapproval to a degree only raised by the imagination, if Joseph continued to display the freedom and independence of his dawning adulthood. No matter that his son had just turned twenty one. The powers of the father would not be denied: his points of anchorage and leverage must remain unchallenged, never weakened. Could a father who knew no bounds, or an interrogator in a closeted police-state, survive and flourish if there was openness, an open society, with the accountability of witnesses?

Mr Salem's mind was filled with fear. Fear of losing control of what he himself had inherited, fear of suffering family disloyalty which seemingly misunderstood this inheritance. Especially from his only son upon which all farming expectations rested. On hearing Joseph answer with pride, confident that his mind was at last made up, these thoughts of losing control and disloyalty flashed before the man's eyes. Yet, even if his son had made up his mind, did that really matter? An unacceptable decision which excluded the farm inheritance could simply usher in the secret 'art' of coercion. Isolated behind a closed door, Joseph would just have to be pushed, prodded, persuaded in the right direction. In this interval of silent strategizing, the conversation had ended.

Joseph piped in: "I also have other reasons to mention the *kopje*." Turning towards his father's profile, he turned towards a sullenness, but was determined to keep the conversation light. "Dad, ever wondered what the *kopje* might mean to me?" The atmosphere at this moment, however, was vacuous, quite without interest. "Do you believe in talents, Dad? As you have been a successful farmer yourself. Ever thought that maybe I am a born biologist? The *kopje* is proof of it."

Like an African toad the man sat expressionless among the cushions, almost unhearing, stationary, his half-seated posture pushing his chin and beard against his chest in a kind of hairy double chin.

"But you just said the *kopje* was for childhood things," he replied.

Joseph's eyes widened: "Absolutely not! I didn't mean it in that way."

"Don't twist your words, son. Say what you mean."

"That's ridiculous," Joseph admitted. "I meant the *kopje* got me interested in biology when I was just a boy. Then my undergraduate studies furthered that interest. Being interested in biology is not a childish thing. At university we have a whole department of scientists teaching, studying and researching biology."

"You know, I never really liked scientists. Did I ever tell you that?" Mr Salem turned his head briefly for the first time. "Look at the world. See how scientists are screwing it up. Technology has a lot to answer for. Animals going extinct."

Joseph looked at him shocked.

"If you look at history," Joseph began defensively, feeling as if a reply was necessary, "scientists themselves seldom abuse science. Because they are seldom the exploiters of their own discoveries. It's the abuse of science by the military, by politicians, by big money from industry and multinational corporations, misusing power and intent on profit that puts this world in the state we find it."

"Pah, politicians are just as bad. I agree," said Mr Salem. "However, think of the merits of being a farmer."

Joseph sighed with exasperation, frustrated at how this conversation was regressing into childish banter. "I can't continue this discussion when such absurd generalizations about farming or being a farmer are bandied around," he said with folded his arms.

"Son, you owe me."

"Perhaps I do. In some ways."

"One hell-of-a-lot you owe me."

"Maybe," Joseph stared ahead at the closed door.

"Why do you abuse me?"

"I'm sorry Dad, that doesn't make sense."

"Give me your hand," Mr Salem re-extended his muscular arm and tapped the wheelchair repeatedly, impatiently with a stubby finger. Joseph stared at his own swollen knuckles which by now were throbbing red and sore, the fingers difficult to bend. He shook his head.

"No, I won't. You hurt me earlier on."

"What rubbish!" The man guffawed with laughter, then sunk into a deadly serious expression. "I thought you professed to being a man? You're twenty one now. You abuse me Joseph: you know how disadvantaged by comparison I am on this bed, yet you don't even have the courage or courtesy to extend your hand to me as a gesture to your father."

Joseph extended his hand limply.

"That's better." Mr Salem smiled through his beard. "I'm glad you agree you owe me something."

It is important for the reader to understand at this point the extent to which this whole conversation was being conducted with a minimum of dramatization. A minimum of observable emotions by someone who may be looking in. Any onlooker out of ears range would consider this exchange between father and son to be quite normal, the holding of hands even to be heart-warming. Not unlike the handshakes and unheard dialogue of two sparring politicians relayed from the distance of a television screen. When one of the combatants at least is well versed in the art of cover up, the art of suppressing his emotions, so that the other must follow: any hostile flare-up of feelings, any outward aggression is tactically weighed up to be used only at the right time in the right place.

"My concern and interest," Mr Salem said quite unashamedly, "lies with the farm inheritance. That should be obvious and needs no further explanation. What you absolutely owe me, Joseph, and don't ever forget it, is this family's share of the farm. Hell, son, do you realize how many people would give their eye-teeth for a piece of *Salemkop*!" He gently stroked Joseph's red knuckles.

"Dad, you presume too much."

"Why do you say that?"

"I haven't mentioned what my decision is yet."

There was another silence. Punctuated by a short bout of coughing.

"Oh… well good. Then let's say this was just another pep talk." Mr Salem continued to stroke Joseph's knuckles.

Joseph looked down at the red swelling soreness of his hand, then up at the man's unflinching profile, and thought: 'this is how my father pep talks.'

*

Questioning the world on a deep level, and intent on finding solutions, was Joseph's natural ability. Whether others considered him gifted or not, it was simply his way. Joseph lived largely unaware of these questions in his head, unaware how deeply he probed them, unrelentingly, passionately, like a dog unable to relinquish a bone. Joseph's felt magnetized to ask the deeper questions. Fertile in imagination. Open of mind. And much as he relished discovering the unsolved, these questions simply led him to more questions. It was a natural born curiosity. A curiosity expanded through his ability to self-teach as a child, then during his university years this curiosity was honed to a sharper edge.

His enquiring mind sponged information and facts with a powerful memory. Soaking up as many viewpoints as he could find. But like anyone, Joseph viewed the world uniquely, at first believing that others saw things precisely the way he did. Believing he was no different. Uncomprehending that his natural habit of questioning everything had bestowed him with a powerful intellect. But Joseph's intellect was also innocent, naïve in thinking he was equal to others and that others saw him as an equal. To him, his broad view of the world was typical of most other people. Equal to most other people. His intellect he treated like a hobby, and he knew everyone had hobbies. It was just that different people knew different things. But university altered this intellectual view of himself. At university, his intellect was treated as a rarified thing. It was recognized for what it was. Rare and something to be prized.

"Do you realize just how questioning you are, Joseph?" someone at university had once asked him.

He shrugged nonchalantly.

"Do you realize the courage it takes to not develop a habit of living with just a few questions?"

To Joseph, the statement seemed absurd.

"I am who I am," he had admitted, without pride or malice. He certainly never thought of himself as courageous.

Joseph's world view had always differed from his father, but he had never recognized it as an issue. After all, a love of biology was an unusual hobby that few shared. But with the nationwide State of Emergency, the country having entered a low-level civil war, Joseph's naturally enquiring mind no longer restricted itself to biology. Beyond the environmental group he had confided in Buck about and wanted to launch soon, the past few days had taught him considerably more about apartheid.

He always knew that South Africa had isolated itself in the world. But his struggles this weekend in realizing what his talents really meant to him found an anchor in also realizing that the way open and free societies determine what is right and wrong is when a majority of people have the freedom to question for themselves. Indeed, as Simon Thandiwe had explained, an unquestioning leader or government can easily be swayed by their own vision of correctness. That even the perpetrator of crimes against humanity often believes themselves to be right. Believes themselves to be doing the common good. For such perpetrators, said Simon, their strength is gained and their criminal courage is found by abolishing this concept of asking what the majority of other people think.

For Joseph, his trips in search of answers this weekend had broadened his horizons. He had asked many questions of many people. By comparison, Mr Salem had remained laagered in his bedroom believing in the scourge of outside interference. And now, seated next to his father, that mushroom cloud of numbness for Joseph began to envelope his injured hand, easing the pain in his knuckles, leaving just a pulsating heat. His personal search for answers this weekend had made him feel more confident.

"I discussed the inheritance with Uncle Jack," Joseph said, blocking out thoughts of pain in his hand.

"Jack must have influenced you," replied Mr Salem.

"No." Joseph shook his head emphatically. "Actually Uncle Jack was very open and honest about the farm and about signing those papers. It was a great help to talk to him."

"Jack is a grabber. I've told you before! You know that! Remember how he moved my labourers above the road merely a month after my accident. And now he's replanting the pinotage vineyard we had decided to dispense with. Ripping up the hanepoort extension that I had developed. I always thought diversity was unnecessary. Joseph, don't be fooled because he's your uncle. Don't be deceived by his sweet caressing words, especially regarding our share of the farm. Where negotiations in the farm are concerned, Jack is even more suspect. When all is said and done, he's hard in here." Mr

Salem prodded a finger into Joseph's chest. "Hard..." he said, repeating the gesture in an almost primal way.

Joseph flinched and turned his chest away.

"Well, I disagree with you. And now that I'm twenty one, I can legally disagree."

Joseph watched as his hand was again taken up into the man's grip.

"What did Jack tell you?" Mr Salem scowled.

Blinking at the pain returning to his fingers, Joseph's voice became subdued, hesitant. "That regarding the inheritance, there's actually no rush to take it up. The inheritance can be signed anytime... in years if needs be. Because there are no other eligible Salem males."

"Except for him, of course," Mr Salem sniggered cynically.

The room fell back into a silence.

"Son," Mr Salem started up again, his tone now changed to a more genial, persuasive voice. "I apologize for this last outburst..."

Joseph remained quiet. Confused. Scared.

"And I'm sorry for complaining earlier about my afternoon worries locked here in the bedroom." His head turned as he stroked Joseph's hand once more and let it go. A strange sorrow was etched upon his face and in his eyes there was a light twinkle of regret.

"Why do we allow tempers to come between us?" he said, as if admitting this in confidence.

Joseph shrugged.

"You're my son. My only son. The last thing I want to do is turn you against me or against Uncle Jack. I've tried my utmost to protect you. And yes, you're now twenty one!" He smiled. "And I'm proud." Mr Salem shifted his position on the bed. "U-hh, my foot is twisted: can you straighten it out please." Joseph lifted the woolen quilt and straightened the limb with his good hand, then climbed back into the wheelchair.

In silence Joseph pondered the sincerity of these words. He swallowed nervously.

"Dad, do you have Jessica's music?"

"How do you know about that!"

"Does it matter. Have you got it?"

"Don't be insolent." A spray of spittle formed on the man's moustache which he wiped away with a tissue. "Don't mince words with me, Joseph. Tell me how you know about it."

"Jessica was upset. So was mother."

"Pah, so I'm all alone in the world again. With my whole family plotting against me!"

"Dad that's not true. A minute ago you suggested that we control our tempers. You even apologized."

"Do you really expect me to feel apologetic now when you're all ganging up against your father and rebelling?"

"We're concerned: Mother, Jess and myself. Why did you have to take her music book away? What has this farm issue to do with Jessica?"

"My dear boy, I'm hardly answerable to you. I turned twenty one many years ago. Now get yourself out of this room and tell your mother to get in here immediately!"

"Oh, hell!" Joseph muttered under his breath. "Dad, can't you see our reasoning: we're concerned about the well-being of the family. For God's sake don't take it out on Mother. Here, look at my hand..." Joseph held it up, limp and red, unable to move one finger while the other two were swollen and becoming bruised beneath the skin. "Don't you think you've done enough damage already?"

"Joseph! Don't you dare utter the Lord God's name in vain. Don't you dare blaspheme in front of me. As a scholar of the Old Testament, at least I deserve that respect. And how dare you preach to me how I should treat my wife, how I should run my marriage. I'm a man of experience and of discipline," he hissed. "NOW REMOVE YOURSELF!"

- CHAPTER 12 –

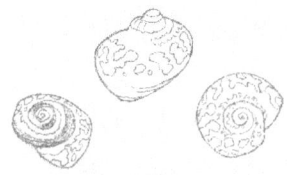

The mushroom cloud of numbness had imploded. The moment Joseph closed the door to his parent's main bedroom, his numbness was left behind in that shadowy room with his father. Flushed away as he exited and none too soon, it was replaced by a gaping hole in his thoughts and his plans. It was an emptiness so unexpected, so filled with a longing for Lexa, for her touch, her gentleness, her feminine words. And fired: 'RAT-TAT-TAT' like a gun with the throbbing of his reddened mangled hand.

If there was any danger of his mother being physically abused, any inkling whatsoever in Joseph's mind of this possibility, he would never have left the house. But he felt that his father's disabled state posed little chance of physical violence against her. Anyway, they were too reliant on one another: his mother and father. With his left arm held limply and painfully at his side Joseph stepped outside into the impending *Kwaaiwater* sunset. The yellow horizon was sprinkling its reddening light onto the ocean in shimmering spangles of gold.

As he headed down to the ocean's water edge, small eddies of froth were being whipped up in the cove. The froth floated delicately in long streamers and in vortices on the swell. All numbness to his senses was gone but that strange inexplicable calm remained: a lightness of being, an empty yearning for the softer feminine touch. To escape from male confrontation. To remove himself from aggression, from his father's anger and intolerance.

So, he jogged slowly with arm held tight against his chest down the steepness of the path onto the beach and stopped when he was knee-deep in the waves. Bending down he plunged his aching hand into the ocean in the way a blacksmith plunges a red-hot horseshoe into a pail of water, expecting a puff of steam to rise from the waves. The hand uncurled like a sea anemone opening, his fingers all softened with the swelling just floated around like puffy tentacles. The low temperature of the water had a soothing anaesthetic effect. Joseph drew out his hand and wrapped it in a checked handkerchief, the hanky Lexa once had given him as a gift with his initial 'J' embroidered on the corner in black. The unwrapped ends of his fingers had turned blue with the cold. Again, he plunged the whole wrapped hand into the icy water to allow the hanky to absorb the sea coolness, then made his way up the beach.

Lexa's house was a ten-minute walk away. Taking on the uphill and heading towards her house he began to contemplate once more human nature. Contemplating some of the findings in his journal, it struck him again that the most difficult thing to understand about ourselves is the ease with which human nature corrupts itself. The ease with which it can confine itself to narrow-minded limits, if not checked. A weakness (plus a strength) that almost defines us as human beings. Though as in nature, Joseph knew, not even this weakness is immutable – he was sure. Nothing in nature is fixed. Not even the rise of this apartheid state in South Africa could be fixed. The ability for change is as much a human characteristic as it is one of nature, a certainty of nature which has saved us many times from our own history.

Joseph hoped he had begun to break his own narrow-minded limits over this weekend? The limits of his family – in outlook - and being raised at this apartheid time. It was very difficult, however, to analyze objectively his own weaknesses. The twinges of his gut. The fact and the question why he had not yet laid down his decision to his father. As much as it was sometimes far easier to analyze others, like understanding his father's capacity to hurt the family. By understanding his own susceptibilities, he felt he may be able to create more choice for himself. Create a semblance of democracy and balance in his troubled mind. In this respect it had been useful to seek out the few people he could count on: Professor Harris, Uncle Jack, Buck, Lexa, to widen the perspective of his goal. Shortly he arrived at Lexa's home.

"Joseph dear - come in!" Mrs. Du Plessis stood at the paneled front door which she held open. "I hardly expected to see you until your party. Very many happy returns of the day. We're in the lounge," she said in a matter-of-fact way.

He followed Lexa's mother, a portly woman who was jokingly referred to by some *Kwaaiwater* residents as the green lady because invariably she clothed herself in olive knee-length dresses. Her occupation at the Town Council was as Hermanus archivist. Her husband was equally conservative, a bird-faced man who had the inscrutable habit of rushing into his bedroom to put on a formal jacket each time a visitor called at the house: winter or summer, night or day. But as Joseph was by now almost one of the family, she warned her husband by calling: "Its Joseph!" and Mr Du Plessis sat put unjacketed.

The sitting room was modestly decorated with a row of three porcelain ducks flying across the wall. Lexa was standing, she had risen from her seat at this calling out of Joseph's name. Her smile looked wonderful to him. Uplifting. Energizing. Though she made no effort to cross over to him or take his hand in any overt show of emotions. Obviously, she had not told her parents about the change in their relationship. In the language of their conservatism, the subject of lovemaking held too many parental taboos. Mr. Du Plessis sat upright in his chair as though he was the last pillar supporting the last bastion of human dignity.

"Evening Mr. Du Plessis," Joseph greeted him.

The bird-faced man lowered the back page of the sports section of a newspaper he was reading and nodded pleasantly, with a tight-lipped smile. His long features complimented his refined movement and the neat and gentlemanly way in which he conducted himself.

"Lexa love, we've about finished talking haven't we?" The lady in green sat down next to her husband as Lexa agreed with her mother. "Off you two go then," she said as Mr. Du Plessis's beak-nose offered them a heedful salute.

Lexa led the way out of the lounge and in the kitchen she couldn't bear to hold on any longer and turned to hug Joseph. "I missed you today," she whispered in his ear. Pulling back to see his response, her eyes were fired with suggestions of what they had shared last night. "Fancy something to drink?" she asked. With her foot she pushed the kitchen door half-closed to add privacy.

"Thanks, but I've just had a tea."

She poured a tall glass of apple juice for herself, dropped in a few chunks of ice, and sat opposite him at the kitchen table. Noticing the wrapped wet handkerchief he was holding, she frowned. "What's wrong with your hand?" Joseph dismissed the question.

"I've just come back from the farm," he said.

"Oh. How was it?"

"Everything as normal. Actually, I went to confirm my decision in my head and to speak with Uncle Jack."

"So you've decided on it!"

"I've decided," Joseph said soberly.

Her eyes widened. "Joe, you won't be sorry. Not with my support on this." She squeezed his good hand across the table and looked at him with an eagerness, a persuasiveness in her eyes. "It's definitely the right choice for us."

"What RIGHT choice?" Joseph's voice was hard, reproachful. "Those vineyards have no place in my life! I saw the only thing that really meant something to me: THE KOPJE."

She drew her hand away. "Does your father know?"

"Lexa, how could you begin to believe, after what we've shared together outdoors all of these years, that I would throw up my studies in biology?"

"But what about those stomach cramps last night? You obviously realize deep down that you owe your father something. And what about how our relationship has changed!"

"None of these things alter our careers. Anyway, where do you see your drama studies in all this?"

"They're important to me," she confessed. "But they don't take precedence."

"What about your talents, Lexa?"

She shrugged nonchalantly.

"Well, I've got to give biology a chance. To take up the award offered to do the doctorate is an amazing opportunity for me. A unique opportunity. I must grab it. Anyway, in effect I've already said yes to Professor Harris."

She looked down. "Have you told your father?"

"There wasn't a chance to tell him outright this afternoon. Stupidly I mentioned Jessica's music and he ordered me out of his bedroom. But I'm sure he realizes my position now, of that I'm certain."

"What happened with Jessica?" Lexa sat back.

"He confiscated her music in an atrocious mood this morning. He was angry because of the farm inheritance issue. At this moment my mother is getting the brunt of it for mentioning the music book to me."

Lexa's mouth dropped open astonished. "What are you going to do?"

Joseph's attention, however, was distracted by the African gardener outside struggling to push a droning lawnmower between rocks up the slope of their garden. "Look at that," Joseph said as the mower hit a rock and came to an abrupt standstill, the gardener disappearing briefly in a pall of blue smoke and dust from the scattered sand.

She looked out the window. "It's only Ben."

"LOOK. Can you see?"

"What am I supposed to be looking at?"

"CAN'T YOU SEE!"

With some agitation she gazed concentratedly out of the window scanning her vision from side to side, not really seeing anything out of the ordinary. She shrugged her shoulders.

"Right there... LOOK AT BEN. An African worker sweating like a pig to get that noisy contraption up the slope. Only to push it back down again and again and again, all his concentration focused on saving the spinning blade from the rocks and from chewing up his bare feet. For that job he goes home to his shack with a meagre wage. Maybe dreaming about freedom on the way, or some decent working boots, or a better grass mower: trying to forget the impossibility of his situation. Has anyone asked him what he wants to be? I'm sure not. He's completely shackled by apartheid."

Lexa was confused.

"The reality not only for Ben but for everyone in South Africa," Joseph continued intently, "is a struggle for their own talents. A struggle to express themselves. This despicable racial discrimination prevents so many people in this country from realizing their potential. And my struggle against my father is not a whole lot different. Biology means that much to me. It means that much. My life - my university career - is not going to regress into something like that: like Ben's life forever pushing a lawnmower up a hill and then down again. Farming would force that onto me. Farming would force me to deny who I am!"

He glanced out the window again.

"I'm going to give Ben a hand with that lawnmower." Joseph stood up suddenly. "Lexa, I'll see you later at the party."

And as unexpectedly as he had arrived, he was gone. In a manner that if she hadn't seen the expression of sheer determination in his eyes, she would have labelled as rudeness. Unmanageable. Unstoppable. Joseph would not pause for a moment.

She had seen that driven expression in the eyes of some anti-apartheid activists. Not that there were many activists at her conservative university but Lexa had seen those eyes before: obsessed and driven, fired up with anger and seriousness, and their intensity frightened her. The thought sent a shiver like an electric spider shuffling down her spine. She watched Joseph cross the lawn and signal to Ben with a broad and friendly wave of his arm.

But why, she scolded herself inwardly, am I reacting like this? Why am I allowing this to happen? Sinking my claws into him and becoming possessive and irritating. Sentimental. Cloying. That's not how I want it to be. Moreover, it's the worst thing for Joseph. Especially this birthday weekend with the decision he must make. With him desperately trying to etch out his independence, trying to fight for his rights, and then I come along to hijack him with my childish romantic expectations.

The irony which Lexa saw was that she herself, inwardly, resented these very expectations which she had voiced, without even wholeheartedly believing in them. She knew where they stemmed from. They were the values her mother and her father held. They were values that still spoke to her deep down in her soul. After last night's change in her relationship with Joseph, these knee-jerk expectations seemed to creep up on her so thick and fast: expectations that were understandable, but hard to control. She knew she would have to control them. Conquer them. She knew she must try to be honest from the heart. Any relationship with Joseph could never be run on the expectations of her parents.

After all, she thought, I am quite progressive in my outlook. Quite liberal and accepting. Maybe my career ambitions are not as determined as those eyes Joseph flashed at me, but I have already shown independence in choosing drama. Joseph said he admired my strong show of independence. And eventually, she remembered, after much consideration and discussion so did her parents: although for them they were ephemeral female options, never something to be taken simply for herself. Lexa knew she would have to keep fighting for herself every step of the way.

She closed her eyes as if in prayer.

'Oh, but how I need to remember this!' she thought. Joseph is not wrong or any less eligible as a partner if he chooses not to inherit the farm. Yes, she realized, I've got to put the record straight with him. Beyond the subtle parental expectations that creep up on me, I've got to free myself from the guilt of choosing this independence. At the party I must tell Joseph this. That like him, I will free myself. Cast off this childish behaviour and these parental fetters, and quickly put the record straight.

*

Ben, the gardener, gratefully accepted Joseph's help. As Joseph reached the top of the slope of the garden pushing the lawnmower, to the amusement of Ben, the gardener tipped his straw hat and beamed a broad, white-toothed smile of thanks, as Joseph departed. Lexa had failed to fill the empty longing which Joseph still felt, a gaping hole of aloneness. To some extent he felt desertion. Only the outdoors as ever could assure a certain calm and connectedness with himself, a spirituality heightened by the setting sun. Beyond the Du Plessis property the climb up the slope steepened until he reached the elevated contour path behind Lexa's suburb of houses. The track wound ahead along the mountain's contour high above the level of roofs and lightning conductors from thatched roof cottages, all the way to Fernkloof Nature

Reserve: a small mountain wilderness where he knew baboons lived. Leopards were much rarer.

Strolling easy along the flatness of the path Joseph contemplated how much he had learned this weekend about human nature. Where, among all the events, did the simplicity of science lie in explaining human nature? The logic of science that is the key to scientific certainty? It was one thing to try to explain how consciousness had evolved, but quite another to try to track the inter-connectedness of consciousness between people. Where was the measure of certainty in human nature in the events of this weekend? The emotions he had experienced were so convoluted, the relationships so complex leading one into the other without beginning or yet any end.

Surely there must be an end to this conflict? Surely some sense of conclusion and progress since the goal of his career was now clearly defined. And with this final decision he felt the calmness of nature without a single stomach cramp, not even a murmur! Except the tingling sensation along his spine which was perhaps not unlike that baboon's raised hackles in the poster, he thought. Because now the encounter was undoubtedly in full cry. His bandaged hand was a painful reminder of that. Yes, there must be an end, and that finality felt very near.

Rounding the next sharp bend on the mountain track Joseph became absorbed into shadows cast by the low-dipping sun, a shady space dark and cool and dense with fronds of ferns heavy on the ground, wild asparagus plants casting their runners far and wide. Water trickled over moss-covered rocks and then across the width of the path. It was an enjoyable level walk, and he strode out watching for loose wet pebbles which may sprain an ankle. A burst of purple flowering heath flowed up the slope rejoining across the track.

There was a faint sound of barking in the distance. Joseph scanned across at the towering sandstone ridge, but there were too many shadows and jagged outcrops to clearly distinguish the movements of any animals above. Was it the barking of a baboon troop? Upwards he squinted at the rocky ledges while continuing to walk along the path. When the noise appeared to be getting louder, he located its source this time out in front of him around one or two more bends. Joseph stopped to listen, a little nervous now. Expectant of something wild, dangerous, while he was quite alone here and unprotected. Until, accompanying the barks, rising high and melodious, he identified the pulsed strains of a tune being played on a harmonica. It was Buck with Randy!

The dog came sniffing around a rocky bend at a place where pink *Watsonia* flowers were growing in pockets on the crags above the path. The dog stopped and cocked its head on seeing another person, recognized Joseph immediately, and tail wagged in greeting. Buck came strolling into view with both hands to his mouth, his shoulders swaying to the rhythm. He stopped and lowered the harmonica from his lips.

"Randy, look who's here."

The Labrador circled, sniffing at Joseph's trousers.

"Good to see you, Buck. I hardly expected it."

The old fisherman looked as windblown as usual with his white hair woolly like an old English sheep dog. "*Muishond-kind*, you look much calmer than yesterday," said Buck, standing in the orange glow of the sunset.

Joseph smiled. "I feel a whole lot better. I've finally made my decision." For him the smile that now came easy seemed to stretch back the years. Having been walking with his painful hand cradled against his abdomen, he lowered the arm unobtrusively to his side. "Thanks for your letter, Buck. I agree that apartheid is THE problem." At this statement, the old man's eyes glinted with mischief.

"My *kindjie*, whether you realize it or not you actually made your decision years ago. I knew exactly what you were the first day you brought that mongoose to me. Aged ten. Am I right or am I not?"

Kicking at a pebble, Joseph nodded.

"Even yesterday when we went to visit Simon Thandiwe, you knew deep down what you wanted, but you needed to push that knowledge to the surface. You were searching incredibly hard to get outside of the fence which other people have put around you. And now you're free. Now you can look in. But *muishond-kind*, we will always

294

need to be on guard for new people who want to tell us what we should be, who are ready to erect brand new fences around us. Some may even want to ring-fence us with reinforced-steel."

Joseph laughed. "Buck, you know me too well."

The old fisherman let out a guffaw, a great booming cachinnation that resounded through his body in shaking bursts, his beard heaving up and down to the movement of his mouth. Both watched the other laughing until tears of relief filled each of their eyes and they both collapsed into a sitting position on the path, gasping luxuriously for air. Then together, side-by-side, they shared a quiet communion by watching the slow change of sunset colours on the horizon. Joseph sat leaning his shoulder up against Buck.

"Nature has been so consistently good to me," said Joseph, pushing Randy away from licking his face.

"A-hhh, that is true... very true." The fisherman turned his head with a piercing seriousness in his eyes. "*Muishond-kind*, I have been rescued so many times from my own folly by just sitting quietly and letting this mother nature of ours wash away all my adult presumptions."

"Buck, what do you feel about being completely alone. Is it as bad as the alienation apartheid brings?"

"Let me tell you," he began, "that living successfully alone is a great art. A loner who has lots of friends but who is truly satisfied to be alone is a supreme artist: a Pablo

Picasso, a Vincent van Gogh, a Michelangelo. But whoever heard of a great artist reaching that degree of greatness without a great capacity for experience, without having lived a great deal and being interested in and having painted very many subjects."

Joseph threw a puzzled look at the old man.

"You see a person must know the reality of what they are missing in other people. A loner should already have experienced life and love deeply and fully so that when he is tempted, he is familiar with the taste of that temptation. Familiar with both the good and bad tastes of reality so that when he is alone his fantasies are real, not abnormally good or ridiculously bad. Fantasy mustn't be allowed to take hold of your sanity which, when you are alone, it can very easily do."

"And you Buck?"

He giggled. "Once I was a terrible womanizer. A-hhh, I was..." he sighed deeply. "Those were the days. Ja, my *seuntjie*, and I was married to a wonderful woman for twenty years who produced two fine sons. One son also became a fisherman, but he was lost in high seas presumed dead. Years ago, my wife also passed on: good woman that she was. The other son is upcountry somewhere, politically active, I haven't seen him for years, but he still corresponds now and again." It was the first time ever that Joseph saw the man dab tears from his eyes, a gentle silhouetted touching of his lined face with a frown deeply engraved on his forehead. "But now it's Randy and me," he said. "We have taken up our places on the south coast like Vincent van Gogh and the wandering albatross."

Buck clicked his fingers to call Randy.

"*Seuntjie,* if this dog and me are to be at your party tonight we must go now to collect and put away my fishnets. It's getting late," he admitted, pointing at the darkening sky.

Together they stood up.

"I'll need your moral support tonight," Joseph said.

"What… at the party?" Buck joked. "Me? I only give immoral support at parties!" Playfully he raised his eyebrows sporting a mischievous smile while punching Joseph on the good shoulder.

"My father still has to be told of my decision."

"You mean you haven't broken it to him yet!"

"Genuinely I tried to. I almost found the chance until he went off on a tirade and ordered me out of his bedroom."

Buck whistled and rolled his eyes. "So tonight, the shit hits the fan."

"Not if I have my way it won't."

"Then I'll stand by you *muishond-kind* because we both believe that your decision is right." Buck raised an index finger. "You have got to plan your words well, know what your strategy will be: it needs to be sharpened to a fine razor-edge because *seuntjie*, you must be ready. You must be absolutely ready. This time his tactics might be far worse than simply ordering you out of his bedroom. With due respect I must say that your father once revealed to me a vicious and ruthless manner. And only a great determination in what you want will break through. But if by some unexpected means he slips by your defences, then you can count on me to be behind you as reinforcement."

"It's fair to be a biologist isn't it, Buck?"

"It's a birthright. It's your birthright. You have never hesitated before in playing out that birthright so don't you dare hesitate now in standing up and defending it. This obstacle that is denying you must be put in its place as it is hurting you. How dare you let anyone – even your father - play God with you, my *kindjie*." Again, Buck looked up at the darkening sky. "There is not much time left for you to think how to sharpen your wits. Hone your courage. You must do it quickly on the way home."

Joseph felt his blood rising.

"And *muishond-kind*..." Buck paused before walking off. "When the time comes, wait for me first before you commence."

- CHAPTER 13 –

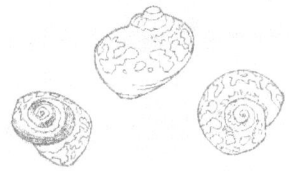

The moon was slow to rise over the Kleinrivier mountains behind *Kwaaiwater*. Against the night sky, the Salem property of *Rus en Vrede* blended almost completely into the coastal shrubbery as Joseph approached the property from the rear. Opening the backyard gate, he disturbed a cricket chirping somewhere hidden in the grass and waited quietly, alert for any human sounds or shapes. He crossed the lawn in deep shadow and entered the house through the kitchen. The interleading passage from the kitchen was deserted and dim but there was a sound of voices in the lounge: his father's distant laughter intermixed with the murmur of other people which seemed to be coming from the porch. Joseph felt like a burglar casing out his own home. But the strategy was set.

He stepped into the dim interleading passageway. Five silent strides placed him outside his parent's bedroom, the door was wide open, the room empty and lit by an embroidered bedside lamp. Joseph listened again to ensure no one was coming and closed the door gently behind himself. The double bed on which his father had lain earlier had been re-made by a woman's neat touch. Crossing the length of the room he reached the yellow-wood cabinet but found it locked with its keys removed. Quick, he thought, where are the keys?

Imagining the possibilities of where the keys might be kept, Joseph suddenly paused. From the passageway there emerged the sound of feminine footsteps approaching. Joseph froze where he was standing. He recognized his mother's voice as she passed by the bedroom door and Lexa replying:

"Don't worry Mrs Salem, I can remember their order. Three beers, one brandy and coke, two sherry's and an orange juice."

They walked past the master bedroom without suspicion, not noticing the door was now shut. Joseph sighed with relief as he heard them enter the kitchen. Domestic noises followed: fridge being opened, drinks being poured, the tinkle of ice being loaded into glasses. When their footsteps returned his mother, who walked in the lead, PAUSED OUTSIDE THE DOOR! In haste Joseph tiptoed delicately to the far side of the bed and dropped to his knees, head held low to the antique carpet.

"That's funny!" Mrs Salem's voice was muffled behind the door. "I'm sure I left this bedroom door open. Do me a favour will you, Lexa, when we've taken these drinks in: come back and check for an open window. The wind is picking up so the door must have slammed. There's a doorstop inside which you can fix in place."

Fiercely, Joseph raced around the bed. Opening his father's bedside drawer, he thankfully found the set of cabinet keys. Lexa would return in a matter of seconds. I have to recover Jessica's music book, he told himself, his thoughts clamouring for time. Then I must get out of here, he knew. The yellow-wood cabinet had four drawer compartments and there were four keys on the set each with notches etched on their respective handles for identification. Starting at the top compartment: the key with one notch fitted, he discovered no books in the drawer and duly closed and locked it. The next compartment down was filled with his mother's trinkets. Skipping a drawer he opened the last drawer on the off-chance:

THERE IT WAS!

Joseph extracted the book of sheet-music with his uninjured hand, but was unable to lock the cabinet without putting the music down on the floor. Bending to pick the book of music up again he heard Lexa's footsteps returning.

Stand up and act natural, he commanded himself. There was enough time to return the keys to the bedside drawer as the bedroom door opened. Joseph spun round, his mouth dry, while he felt quite defenceless in his role as burglar: injured hand limp and useless, the other hand clamped around the music book. But why do I feel guilty, he asked himself, when this is for Jessica's sake? Lexa walked in casually expecting no-one. The vision of Joseph hunched over the music book and staring ahead in surprise prompted her to half turn as if about to run away before she recognized who it was. For a minute she was speechless as her breathing heaved.

"God, you gave me a fright!" she gasped.

Joseph straightened up as he glanced back at the bedroom curtains, at the same time tucking the music book under his shirt. "I've already checked for any open windows," he said nonchalantly, offering this as a quick excuse. "Everything is fine in here." Then he walked past her with added purpose in his step. "I'm sorry if I startled you, Lexa. But I ought to go up to my room to get showered and dressed."

"Wait - why were you..."

And he disappeared quickly back into the dimness of the interleading passageway.

*

The shower felt warm, satiating on his skin, and in the rising steam Joseph found peace enough to hum to himself. Phase one of his plan was complete. A calm had returned in knowing there was some progress. And furthermore, what was about to be done was undoubtedly the only option possible for him now. In his mind and heart, it had a greater truth, a saving grace about it that was genuine. So far, the plan was unfolding smoothly and without a hitch.

Joseph uncoiled the hanky from around his injured hand. Some of his fingers had become useless, three were immovable due to the swelling, too painful to be extended to any degree. And they had taken on a hideous grey-blue colour like an overripe bunch of bananas, swollen and curved in a half-closed clench. The hand felt less painful when cool and bandaged so he rinsed the checked hanky in cold water and rewrapped it, securing the end with a safety pin. Naked down to his torso, dressed only in a towel around his waist, Joseph stepped from the adjoining bathroom into his bedroom and unexpectedly into the presence of Lexa. She was standing beside the dissecting table at the far side of the room peering into jars of preserved specimens and inspecting the Malawi cichlids in the fish tank. She turned to face him as he entered:

"Joe, I need to talk with you."

"If you're not bothered watching me dress."

She smiled coyly and sat on the bed. "I've been acting like a silly child," she said.

Joseph offered no comment. He opened the wardrobe.

303

"I'm sorry but I wasn't myself down at the beach or when you visited my house this afternoon. It was a childish way to behave, throwing those expectations at you, and my fault for getting so emotional about our relationship. I hardly recognized the things I said."

"Lexa, I'm not judging you."

"You probably realized that, more than anything, I was voicing my own parents' expectations."

He nodded and smiled.

"But by now you ought to know that mine are different."

"Look, we were both overwrought," Joseph admitted. "I left your house in the hope of preventing another confrontation, especially trying to avoid one involving you. I knew you would quickly see sense in those expectations of me, so I left you up to it. And now you have come around to them and I'm glad of that." He dropped the towel to step into a pair of underpants. Lexa turned her eyes shyly away but not before seeing his manhood and feeling the ambivalence of a young inexperienced woman towards it: slightly scared, yet fascinated.

"Will you give me a hug," she asked.

Joseph drew up his trousers, walked over and sat down beside her. She placed her head at rest upon his bare chest, sighing as he held her and gently twirled his hand through her hair. He noticed how stunning she looked for the party. It was the first time he really noticed.

"So will you tell me why you were in your parent's bedroom with the door closed. You scared me."

"I had no intention of scaring anyone," he said, walking over to the wardrobe for a shirt. "I was searching for Jessica's music book. I've taken it back."

"You did what?"

"It was necessary under the circumstances."

"Does she know?"

"That's immaterial at this stage." Delicately Joseph inched his bandaged hand down through the shirt sleeve, encountering some trouble in getting his whole arm through. "I'll need Jessica's music later for when I talk to Dad. To give me extra leverage, for added persuasion."

"You're going to tell him at the party?"

"There's no alternative. There is no other time except just before he retires to bed. He must be told tonight. Buck has offered to stand by me in support."

Lexa felt a surge of disappointment. It had nothing to do with Joseph at long last telling his father, but instead it was the fact that she was being excluded, left out, abandoned she felt at this critical moment in his decision, after having offered him her own personal advice last night. Inwardly she again regretted having thrown those childish expectations at him, for having lost his trust at this time. It seemed to her now that only the right words would help.

"You also have my full support Joe," she said plainly. "I too would like to be there when you tell him." She waited, watching his profile, unsure how he would take this. The line of his jaw fluttered with a twitch of muscle movement. Joseph buttoned his shirt. He did not reply. It seemed he was caught between saying yes and no.

*

By half past nine all guests invited by Joseph's parents had been personally greeted, primed with a drink, and the gift stage of the party was over. Gifts lay opened on one of the lounge settees, nestling in crumpled wrappers. When opening each gift, Joseph had aroused, one by one, the applause of these elderly partygoers. Their toned-down gift wrappings reflected the conservative taste of these guests and set a definite standard of responsibility in the choice of presents given to Joseph. On the porch afterwards they sang a formal:

> "For he's a jolly good fellow
> For he's a jolly good fellow
> He's got the key to the door
> never been twenty-one before..."

This formal singsong echoed into the African night, contrasting starkly with the uncontrolled pounding of the surf outside on the rocks. The people and their voices

306

were stiff and upright in their colonial formality. After the gift opening 'ceremony', Joseph made every effort to spend time talking with his mother and father, thanking their friends and Lexa's family, giving each guest adequate attention. He knew that the Hermanus *Coloureds* would arrive later with their musical instruments. In the meantime, the party was a small sober gathering consisting of familiar and expectant faces that stamped the warm night with a mark of 21st birthday solemnity. Elderly faces that gave substance to the kind words written in their half-dozen birthday cards. Phrases such as:

'Now you're an adult, you have the key to the door!'

Joseph expected two different parties. Therefore, this first formal start came as no surprise. But he did smile inwardly at the absurdity of receiving birthday cards which declared with official approval in bold cursive gold and silver script that suddenly he had metamorphosed into adulthood. What an odd thing to celebrate, he thought? But then privately, when he considered more deeply the extent of his travels this weekend, how he had desperately raced against time to make his decision, this official idea of his 21st birthday celebration no longer felt as absurd. Few in this audience knew how far he had travelled this weekend: in mileage and in his head. But he knew the distance he had travelled, and it made him secretly proud. Yes, indeed, maybe I have metamorphosed!

Perhaps becoming an adult can happen just that suddenly, Joseph contemplated? Maybe there's nothing absurd about it. But he knew that being a twenty-one-year-old had little to do with being able to watch age-restricted films, or being eligible to sign an inheritance. Being twenty one is knowing who you are! Really knowing! Inwardly, as he glanced around at his guests, he pondered who in the audience was a really good example of being an adult, a really good example of knowing who they were? Perhaps I've grown up this weekend more than any of them realize. If they knew my plans, rather than expecting me to accept the farm inheritance, to be the next proprietor of *Salemkop* wine farm, what would they think now? Would they ask for their farm related birthday gifts back when they realize I'm not choosing that particular *key to the door*?

In good humour though Joseph made light of this formal part of the occasion. It wasn't easy, however, wearing a starched shirt in the humidity of this African night, buttoned to the collar, as he had been told to do, and locked into a jacket. He had rejected wearing a tie. For a less formal celebration he and Lexa would have to wait for the arrival of the Hermanus *Coloureds*. Joseph loosened his collar to ease the constriction around his neck with the forefinger of his good hand. Then he stood up and cleared his throat to speak:

"*Aandag asseblief*... Could I have your attention!" The familiar banter subsided as eyes focused onto Joseph.

"Can I extend a big welcome to all our friends and family on this my 21st." To oblige his parents' guests, Joseph repeated this greeting in Afrikaans. "As you can see scattered around the porch on side-tables is a range of snacks and wine, while beer and soft drinks can be found in the kitchen. In fifteen minutes, a cold buffet will be served followed by dessert." The audience let out a low murmur of approval. "Later on, I'm expecting another group of friends and we've planned a barbecue outside for them. Feel free, please, to join us if you will after this buffet..."

At that moment Lexa walked in from the lounge to offload a tray of drinks, then she leaned up against the porch sea windows. Joseph smiled at her and continued: "It was suggested I introduce you to one another, but I think that is quite unnecessary here in *Kwaaiwater*." Another murmur resounded - of amusement this time as people glanced around knowingly at each other. "Before I thank each of you individually for your gifts can we all show our appreciation to my mother, to Jessica, to Lexa, and to our maid Elsie, for preparing this magnificent spread." There was a round of applause as Mr. Du Plessis nodded his beak-nose in agreement.

Turning to these guests one by one, Joseph then thanked them for their gifts. Although absent, Uncle Jack and his family had sent with their birthday card an expensive writing-set together with a case of *Salemkop* hanepoort wine. Mrs Du Plessis wearing an unusually vibrant green outfit almost blended into a giant sword fern hanging in a basket beside her and she and her husband, sitting isolated, had given Joseph an all-weather anorak. The Rossouw's - Gert and Sannie, and the Le Grange's - Johannes and Elize, were both Afrikaans farming families who sat in a tight and familiar circle around Mr Salem. When the time came to thank his father for the gift of two books, Joseph spoke to the whole circle. Because obviously both Gert and Johannes had been primed in their choice of birthday gift: together they had bought Joseph a one-year subscription to South Africa's top agricultural journal. Mr Salem in his effort towards this end had chosen the latest editions of two major farming reference works: *'Comprehensive South African Viticulture'* and *'Scientific Farm Management'*, books quite the envy of all three farmers here tonight.

Continuing this round of thanks Joseph maneuvered his way delicately with a slow deliberate diplomacy around the subject of the farm, his future, and the relevance of these gifts. To the guests listening to him farming was a family issue. None of them really knew Joseph personally except through the lens of his father. Therefore, he was not obliged, he felt, to reveal anything of his future plans to such distant and expectant family friends. Anyway, Gert and Johannes probably had purchased their gift quite oblivious of any lurking implications: theirs was most likely a naive kindness at which Joseph aimed his praise.

However, the choice of personal gift was very different where his father was concerned because there was now an awareness on both sides of the underlying conflict involved. Joseph's voice strained as he caught his father's attention: twice stopping to lower his eyes, feeling an awkwardness at the lie he must act out for these people. Still, he continued, commending his father in public on the usefulness of the books. He hoped his voice didn't sound too drained of its sincerity.

Mrs Salem, however, seemed quite alienated out there watching Joseph. The helplessness of his mother was obvious, as she recognized the strain on him as often only a mother can: her face confused with frustration, with a bewildered caring expression. During one of Joseph's difficult silences, she moved over to sit with Lexa's mother almost as if the Du Plessis' neutral gift of an anorak offered some comfort to her, a resting place for her conscience. Her inward struggle to find some role as mediator was futile now because no longer was there room for her mediation. The issue between father and son had become so cut and dry, so black and white now, the boundaries at last were so sharply defined with dangerously little space left for negotiation. That to her there could be no more severe way of realizing Joseph's entrance into adulthood than by seeing her son and husband squaring up to each other as men.

She knew that her husband's 21st birthday gift presentation meant that Joseph in essence held the next move. As much as she had come to realize that even her symbolic gestures as mediator-in-waiting were futile and had finally forced her out into the audience to watch this domestic battle as an onlooker from a distance. For Joseph this fact meant that he could breathe easier knowing that his mother was out of the fray. Out of harm's way. He even wanted to push her further onto safer ground, liberate her further from this destructive family warfare. In wanting to acknowledge her important position in the audience now, Joseph turned to her.

"And for you Mother, a last word, but by no means the least. The cake you baked me this afternoon made me better understand your point of view: the extent of your broad-mindedness and how much you genuinely care." Joseph stressed these words with as much clarity and frankness as he could. He was sure no one else knew what they had shared back at the kitchen table with the black-current tea anyway. Then finally he made a last scan of the faces.

"To everyone," he said, raising his glass. "Please do enjoy yourselves. *Geniet die ete*. Have a wonderful party!"

*

Above the waves and the subdued onshore wind, the strains of Buck's harmonica, a crooning ode to the evening, preceded the arrival of the Hermanus *Coloureds* at half past ten. Buck was heading a group of twelve friends that reached the seaward perimeter of *Rus en Vrede* as the old fisherman broke into that compelling tune:

'I was born under a wandering star.'

But it was not from strains of this tune that Joseph first ascertained their arrival, nor the other music accompanying the group. It was instead the expression of alarm on Gert Rossouw's face. Gert was standing on the porch facing the sea windows spooning down a bowl of trifle and talking to Joseph about his expected grape harvest when he stopped open-mouthed in mid-sentence. No word was uttered to conclude his sentence, but his look was one of astonishment and glassy-eyed severity. An expression of disbelief aimed out of the window at *Coloureds* who were actually walking up the length of the garden towards the party as if invited guests. Gert immediately glanced around to locate his wife as Joseph stepped outside to greet them.

"Welcome... welcome!" Joseph extended his arms.

The group encircled him. They were wearing bright colours and smiles on their faces lit by the orange flames of the barbecue fires. Soppie had surpassed himself in wearing his Minstrel Carnival outfit of garish green, red

313

and purple stripes, his boater hat worn rakishly at an angle and a banjo strapped across his shoulder. He stepped forward in a happy frame of mind. Taking a low dramatic bow in front of Joseph he swept his arm in a broad arc to encompass the group and shouted:

"Congratulations, Josef, from us to you!"

One by one the group lined up to personally wish Joseph on his birthday, some relaying the greetings of others unable to be here tonight. Soon they huddled around the fires, a few dancing, clapping their hands. Those twirling round the others egged them on by locking arms alternately with different standing figures. It was a way of stirring up those who had walked a far distance to get here, were out of breath and pausing for a drink. Soppie began strumming on his banjo, raising it above his head, alternately plucking and finger-drumming the instrument while weaving in and out of the others to coax them on, glancing mischievously sideways every now and then with a smile that revealed a toothless grin.

From behind Joseph, Buck came up to whisper in his ear: "Many happy returns *muishond-kind*. I'm at your service whenever you're ready to do it. Simon Thandiwe sends apologies for not coming but he's involved in a crisis meeting back at Crossroads."

Meanwhile Soppie was baiting his audience by bobbing up and down in the middle of a dancing circle which had formed. "Are we ready to wish our host?" he

314

cried, receiving a chorus of approval from a dozen sonorous voices. To free his arms for a jig, Soppie tipped the banjo over his shoulder and lowered it down his back until it rested on the leather strap pointing groundward. He shouted with more verve in his voice:

"I said are we ready?"

"*Ja, Ja...*" The reply got louder, more enthusiastic, vibrant and colourful as the whole circle linked arms on shoulders. Glancing back at the porch Joseph was surprised to see his other guests in the house in a formal row at the windows staring down at them. His father seemed to be glaring with distaste at the scene around the barbecue fires and revealed what looked like embarrassment in his eyes.

"Well let's bring on the dancing girls!" Soppie cried out as the group began to clap to the funky rhythm of Buck's harmonica. Nellie and Rosina linked arms with Joseph making him part of the circle, while Katrina pinned onto the pocket of Joseph's shirt a red and yellow rosette with the insignia: 'THE HERMANUS MINSTRELS'.

"At twenty one, Josef, you are now one of us!" said Katrina. These words were instantly taken up as chants from the happy group: "FOR HE'S ONE OF US... HE'S ONE OF US... FOR HE'S JUST A JOLLY GOOD FELLOW..." As Nellie and Rosina wiggled their shoulders and buttocks, Joseph unlinked his good arm to undo the top two buttons of his shirt. In full view of her parents, Lexa boldly crossed the grass to take up hands with Sampie and Nelius in the outer circle. Joseph and the two dancing women were turning anti-clockwise in the middle as the ring of *coloured* faces, a number in Minstrel Carnival make-up and with toothless grins of delight swayed their hips, clapped their hands, and spun in the opposite direction. By the time Soppie entered the circle to raise his arms requesting a moments silence, Buck was red-cheeked from his harmonic exertions. Even Randy barked with excitement.

"Quiet *julle*," Soppie intervened and the leg high-kicking dwindled. "Quiet, *ek se*!" he reiterated. Walking up to Joseph he gestured with an expansive sweep of his arm. "On behalf of all of us, we thank you Josef for this kind invitation. And we congratulate you heartily on becoming a man!" There was a rumble of approval, a consensus of nodding, a few smiles beamed out in the firelight, and as Gideon started to clap, frail old Lizzie broke down in tears. She had known Joseph since his birth. Buck took her in his arms to hold and comfort her.

"Quiet julle!" This time Buck intervened. "Let's hear a word from our *muishond-kind*... except that now he's a *muishond-man*?"

A round of laughter broke.

Joseph left the dancing women, walked over to old Lizzie and planted a kiss on her cheek. Indeed, she had known him longer than anyone else down here at the barbecue. Even longer than Buck. Joseph turned to the others: "Quiet *julle, ek wil praat!*" With these words there was another light-hearted round of applause. Looking back at the porch again only Jessica and her school-friend were visible now standing at the sea-windows looking down on them, giggling in a schoolgirl way.

"*Naand almal... dis goed om julle by my plek to he,*" Joseph spoke above the crackle of the fires. "Friends, thanks for your good wishes: to Katrina who also gave me a scorpion yesterday, to Nellie, Rosina, old Lizzie (Joseph was holding her hand), Boet on the cymbals, Kenny, Nelius I love your yellow top-hat and tails, Robert there with purple make-up, Sampie on the sax, second-banjo Gideon, Soppie of course and our politician Buck and everyone else. Then Joseph drew attention to the rosette which had been pinned to his shirt pocket. "I must admit I've felt like a member of your family for years already." Gideon cried: "But now you're a minstrel!" Gideon started a burst of clapping, but Soppie grabbed him by the arms to subdue him for a moment.

"Tonight, we have two parties," Joseph continued. "Up on the porch some friends of my family have been invited. Down here is a barbecue for us. When the coals are ready, please feel free to put meat which is over there on the fire. We will also be bringing out freshly caught fish, baked potatoes, onions, *mielie pap,* tomato stew and salads. Turning to Buck, Joseph requested his help to

317

carry this food out. Together they left the group and walked side by side up the garden as the dancing banjo music resumed.

At the house someone inside had locked the porch door preventing anyone outside entering from the garden, so Joseph and Buck were forced to circle the property. Looking back Joseph noticed that the *coloured* group had positioned themselves loosely around the two fires and were singing an old slave ship song: *'Daar kom die Alabama'.*

At the front door Joseph was surprised to discover that both the Rossouw family and the Le Grange's were about to depart. His mother was standing at the entrance with them while his father exchanged last words with Johannes.

"A-hh, there he is!" Gert exclaimed in a relieved tone as if he had been searching around for Joseph for some time. "We're off – *ons moet gaan*. Thanks Joe m'boy and best of luck for your farming future." His wife Sannie who stood by his side also offered, with a modest smile and a tilt of the head, her best wishes.

"But you're leaving so early?" Joseph said.

"It's a long drive back," admitted Gert while he glanced with a disapproving eye at Buck wearing an old tweed jacket and who stood beside Joseph. Johannes Le Grange also approached and placed a heavy farmer's hand on Joseph's shoulder. "We must be on our way too," he concurred. "Congratulations again son. I hope you get good use from that journal subscription. I'd give my eye-teeth for those farming books, you know." As the two couples walked off to their cars, Joseph and Buck entered the house passing by Mr Salem in his wheelchair.

*

"My father is going to bed soon," Joseph whispered to Buck across the kitchen table. "Nearly time for phase two! But first I must collect a few documents." The only guests remaining in the house were the Du Plessis' and after his conversation with Lexa earlier on, Joseph felt regret that she must leave with her parents because of her play rehearsal tomorrow morning in Stellenbosch. It meant she could neither witness him breaking the doctoral news to his father, nor would she immediately know the result of the confrontation. Mrs Salem entered the kitchen as Joseph and Buck loaded salads and onions, potatoes, and gutted fish onto trays.

"Can I help you both with that," she offered. "Oh…" She turned to Buck, smiling. "I haven't yet said hello. How are you keeping, Buck?"

"Just fine Mrs Salem. A mighty worthy party this."

She nodded and giggled at his enthusiasm. Then she picked up a tray of food to lead the way. In the lounge they passed by Mrs Du Plessis who signalled Joseph to call Lexa in from outside.

"It's half past ten," the woman-in-green mouthed and pointed at her wristwatch.

After Mr Salem indicated that the porch door key had been safety put on the mantelpiece, they were able to exit into the joyful party atmosphere outside. Metal grids were laid over the barbecue fires and Mrs Salem, Buck and Joseph were met by an exhortation of approval. Even old stubborn Gideon stopped a fast-twisting jive in order to peer and chuckle into one of the salad bowls.

En route back to the kitchen Joseph detoured quickly upstairs to his bedroom. Jessica's sheet-music book and his research journal were placed into a briefcase, he had decided, as ammunition to persuade his father. Feeling confident with this strategic choice, Joseph gave the briefcase a reassuring tap. By his bed in front of the leopard-baboon poster he paused: a butterfly feeling washed lightly through his stomach, then that soothing unfathomable calmness again. He descended the stairs slowly, deliberately, with concentration, until at the bottom he was intercepted by Lexa in the entrance hall.

"Joe... I'm sorry to leave this early. It's really hotting up out there," she said laughing and kissed him goodbye. Holding up her crossed fingers, she winked. "Good luck with your father. Phone me tomorrow afternoon with your news, and we'll finalize arrangements for a ticket to my play on Wednesday."

After the Du Plessis' family departed, Joseph found himself suddenly alone in the kitchen holding the briefcase. His mother and Buck were somewhere in transit carrying food. An eerie squeaking of Mr Salem's wheelchair issued down the passageway as his father turned into the master bedroom. C'mon Buck, Joseph thought with some urgency, we must do it now! As he waited for Buck, the seconds seemed to stretch out forever into minutes, until at last his mother's voice was heard with Buck in tow. They entered the kitchen each carrying an empty tray.

"That's everything taken outside I think," Mrs Salem said opening the fridge. "One last check!" Joseph discreetly caught Buck's attention by making a repeated circle in the air with his index finger to signify the moving wheels of the wheelchair, then pointed in the direction of his father's bedroom. Buck nodded affirmation. Both now understood that the time had surely arrived.

"Yes, that's all of it." Mrs Salem confirmed and put her tray down on the kitchen table.

Joseph turned to her. "Mother, would you be able to help outside, see that everything runs smoothly, and ask Jessica and her friend to help you. I must talk to Dad again and Buck will be with me." An expression of disquiet touched her face as she glanced alternately at the two men, seeing their seriousness and remembering the blow-up in the bedroom this afternoon for which she had taken an earful from her husband. Joseph went to kiss her reassuringly on the cheek.

"We won't be long," he said. "There's nothing to worry about."

Her departure coincided with the opening of the porch doors and the distant sounds of celebration outside around the barbecue fires in the garden. It meant that they were alone with Mr Salem in the house.

"Follow me, Buck," said Joseph. Clenching the briefcase under his bad arm he strode down the passageway and knocked on the half closed bedroom door. "Dad, its Joe. Do you have a minute to talk?"

Buck whispered to Joseph: "No forced removals!"

Joseph nodded rapidly in agreement.

"Joe, yes son... come in."

Buck boldly entered the room after Joseph and Mr Salem's face paled as his eyes fixed sternly, questioningly onto the fisherman. Mr Salem closed the hardcover book he had been reading.

"What are YOU doing in here?" Mr Salem hissed angrily at the sight of Buck in the privacy of his own bedroom. Seated in his wheelchair under a reading lamp next to the yellow-wood cabinet, he had been paging through one of the new farming book gifts given to Joseph for his birthday. Now Mr Salem's stout hands readied on the wheels of his chair in a posture of intimidation and threat, his torso leaning forwards, as if about to stampede the fisherman out of the bedroom.

Joseph took a step to block his path.

"I invited Buck as my friend into this conversation," Joseph answered, standing his ground as his father shifted his glare of severity onto him. Buck watched with some excitement coursing through his veins as his *muishond-kind* withstood the cold destroying gaze. But it was eventually kindness that broke the eye contact between father and son: Mr Salem lowered his eyes when Joseph put a hand on Buck's shoulder, then drew out the dressing-table chair for him to sit. At this point Joseph had decided he would act out phase two of his plan on his feet. His mobility plus the added height was useful. He placed the briefcase on the bed central to the three of them, with Buck seated on Joseph's left and his father in the wheelchair to his right.

"You still haven't thanked me personally for this!" Mr Salem gestured at the two farming books on his lap, running two stubby fingers over them.

It seemed to Joseph a perfect opening from which to commence his advance. His father had unsheathed the first sword. Really was he waiting for gratitude? Really? Joseph glanced momentarily at Buck who winked back in recognition of this opening and that glint in his eye let loose a strong silent battle-cry.

"Personally, I'm afraid," said Joseph stepping towards the bed, "there is only so much to thank you for."

In silence these words were digested. Mr Salem's posture remained undaunted.

"Just what do you mean by that?" he asked.

Buck's presence in the bedroom seemed to have great influence because Mr Salem, although his face was growing increasingly agitated, was by his usual standards quite composed and still remarkably hesitant at this stage.

"I find your choice of gift a tactless gesture. No...!" Joseph raised his uninjured hand to his father with a finger pointing upwards, as Mr Salem opened his mouth in puzzled protest. "Not from a lack of trying to communicate like an adult with you, I have no choice anymore but to employ your own methods of force to make you listen to me. No...!" Again Joseph raised a finger to him. "With regard to my final career decision this weekend, you have assaulted me physically." He held out the other hand, limp and bandaged. "Apart from the psychological abuse this whole family has had to endure."

"NOW WAIT A MINUTE!" Mr Salem hissed, frowning with a furious air about him.

Joseph unzipped the briefcase and took out Jessica's music book, strode over to the wheelchair, and thrust the book a recognizable distance in front of his father's eyes. "See this!" Joseph tapped hard onto the cover. "If you're a

MAN don't you dare start abusing women, children, your own daughter, your wife, to get at ME!"

Suddenly Mr Salem's face went purple, wild-eyed, he smacked the music book out of Joseph's hand onto the floor, then lurched forward in the wheelchair. Buck was on his feet in a flash. He pushed the bedroom door closed, secured it at the lock, and threw the key over to Joseph. The two of them towering in standing positions over the wheelchair brought Mr Salem to his senses. Now both Joseph and Buck stood equidistant from him, and their presence was firm, united, unyielding like a front of steel.

"In your mind this weekend, merely your mind," Joseph continued without a quaver in his voice, "you have blocked out to yourself my basic right to make my own career decision. And then in a selfish and quite intolerant way you set about forcibly laying down your rules, defending them at all costs, as though my own will and reasoning doesn't even exist. No...!" Joseph anticipated the man's protest while Buck was quick also to take a step forward.

"WHO THE HELL DO YOU THINK YOU ARE?" Mr Salem spat the words at Buck.

"Sir, I consider myself... A GENTLEMAN." Buck savoured the taste of these last words, speaking them slowly, issuing them in a wise and refined pronunciation.

"WELL GET OUT!"

The fisherman stood his ground even when Mr Salem jolted the wheelchair forward. Without lowering his intelligent eyes, Buck turned them in Joseph's direction. "I am here by the invitation of your son. With due respect Sir, I shall take my call from him."

Joseph confirmed: "Buck stays with me."

"COME HERE JOSEPH!"

"But now for the good news, Dad." Joseph ignored his father's command and instead he watched the man's purpling face, an expression of helplessness in having to confront his very own tactics. "It seems that my scientific hobby as you like to call it..." For visual impact Joseph extracted his bulky journal from the briefcase. "And my efforts in zoology at university are being rewarded."

"WHAT DOES THAT MEAN?"

"It means I have decided to accept a four-year doctoral scholarship in zoology starting in January next year, all expenses paid." With that Buck walked over and shook Joseph's hand vigorously and enclosed him in a bear hug. They both turned to see Mr Salem's eyes glaze over, his face alter a shade of colour, while he muttered silently, incoherently.

"Dad, I know this sounds sudden to you. But will you wish me well?"

There was no acknowledgement.

A vein in the man's forehead pulsed at the surface of the skin, he seemed wrapped in a private dreamworld of his own, his hands cradling the farming books on his blanketed lap.

"Dad... are you alright?"

A tinge of blue was dappling the man's cheeks and his mumbling changed to soft gasps. Then inexplicably he held his breath as if in pain, his head lolling, his face turning bluer, the contorted whimpering gasps became a whispered cry.

"Help me, Joseph. Please, please!"

"He's toying with us?" Buck challenged.

"No... he can't be!" Joseph grabbed his father's stocky wrist, the pulse was regular, but the man was holding breath again and forcing himself to turn blue. "Oh, God no! He's making himself have an attack!" Joseph could feel his own stomach knotting up. He rushed over to the door with his journal still held under the bad arm, rummaged in his shirt pocket for the key, and finally managed to open the door. Standing there leaning up against the doorframe, his head bowed over the scene, his mind spinning trying to make sense of it, his ears filled with his father's groans of pain and breathlessness - Joseph felt dumbfounded. His panic was rising. What should he do? What on earth could either of them do?

"*Muishond-kind*," Buck interjected. "Calm down!"

"Buck, hell man, can't you see what is happening? NO... better not touch him!" Joseph could sense himself on the verge of losing control, losing himself in this irrational

327

caring and panic for his father partly because of what it would mean to the rest of the family if his father did suffer a seizure. As he had done in the waves. As he had done at the accident. He rushed over and loosened the man's collar. Again, the pulse was normal. "I've got to get him help! Quick, you go outside to call my mother..." The weight of Joseph's responsibility to his father was becoming overwhelming as the man held breath again and again. "Quick, call my mother, Buck! I'm going off to get some professional help!"

And in this state, with a badly injured hand, his journal pressed tight under his arm, Joseph sped through the dimness of the passageway out into the moonlit night.

- CHAPTER 14 –

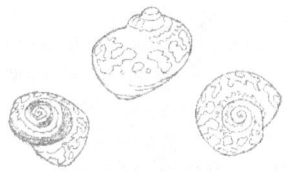

Lexa woke to the ringing of a doorbell. The ringing persisted, over and over and over, as she pulled herself up, half-propped her torso on a continental pillow, feeling bleary eyed with a light headache throbbing at her temples and forehead. She felt disoriented. As if she were floating outside her body. Massaging her temples, she closed her mouth to find that there was a moist patch of dribble on the pillow which had been resting against her cheek. Her neck was stiff and sore, her vision blurred. She blinked repeatedly to clear her sight.

The doorbell continued its urgent ringing downstairs. Lexa looked up dreamily at the stuffed albatross dangling from the ceiling, then at the model of the pterodactyl suspended with fishing tackle in a steep dive above the bed. Mice squeaked in their cages beneath the laboratory bench with its microscope on top, as her vision cleared and took in the shape of a computer and printer, plant presses, bottled specimens all around... Then she remembered where she was.

It was Joseph's bedroom.

Events flooded back which seemed more dream-like than real. The cliff-top accident and Joseph's mysterious disappearance, the blood-stained rock and the trail of white pages scattered along the shore. The subsequent police search. She recalled Sergeant Malan's attempts to extract information from her in that police station cubicle. On the bedside table the clock indicated the time to be half past one in the afternoon: five hours had elapsed since she fell asleep on Joseph's bed. No wonder she still felt bone-achingly tired. She had barely slept last night after raising the alarm around Joseph's disappearance. Now there were muffled sounds of conversation downstairs from the person who had finally opened the front door. Suddenly Mrs Salem cried:

"He's ALIVE! They've found Joseph!"

Lexa pistoned off the bed with her heartbeat pulsing in her ears. In nothing more than a reflex action she found herself careering down the stairs. Mr Salem was wheeling himself along the interleading passageway towards the entrance to hear this unbelievable news when Lexa's feet thudded down to ground level some distance in front of him. Mrs Salem was clutching onto the hand of the policeman at the front door, half beside herself with emotion, pleading with the uniformed man to give her more exact details. The young officer raised his hands in exclamation:

"*Jammer Mevrou.* I'm sorry but I really don't know anything else about the discovery."

The overwrought mother who would not let go of her grip on the policeman was startled by Lexa's bounding intrusion.

"Where is Joseph?" Lexa questioned the policeman directly. "What have you done with him?"

Mr Salem arrived on this scene at the entrance hall.

"I can only repeat what I already told you," admitted the officer in a strong Afrikaans accent. "Joseph Salem was found conscious but in shock although not seriously injured, forty minutes ago. He has been admitted to the Hermanus clinic."

"Drive us there!" Lexa insisted.

"Sorry madam but there are a number of police dogs in the back of my pickup. I have space to transport only one person in the passenger seat."

Mrs Salem became angry. She turned to her husband. "Harry, should I go?"

Looking up at the policeman, Mr Salem cleared his throat to speak: "You say our son is definitely alright. Not critically injured or anything like that?"

The officer nodded affirmation.

Mr Salem considered how his wife was reacting to this news, confused and veering from panic to anger. Lexa standing beside him seemed in a more able state of mind to travel alone with the policeman. "You better go, Lexa," Mr Salem decided. "Phone us immediately from the hospital once you've visited Joseph."

"But Harry, Joe may need me!" Mrs Salem pleaded. She got angrier. "Where is that Sergeant Malan!" she insisted. "Why hasn't he supplied us with proper transport?"

The policeman stepped back awkwardly in embarrassment.

"*Mevrou* Salem," he said, smiling sheepishly as if his mother were scolding him. "Sergeant Malan sends his sincere apologies to you for not coming to tell you himself but unfortunately he is too busy involved with the Buck Williams case, because there have been certain complications."

Lexa's expression changed to disquiet. The truncheon episode she had witnessed this morning in the police station flashed into her mind. Mr Salem shifted in his wheelchair, glancing out of the door beyond the overweight figure of the policeman who was blocking much of the view.

"I requested the Buck Williams enquiry," Mr Salem admitted.

"You, Harry?" his wife asked in puzzlement. "Why didn't you tell me this!" In her surprise she remained quite oblivious to the enormity of her husband's role in initiating police action against Buck. Lexa, meanwhile, visibly paled at his statement.

"As soon as we can manage it Ma'am," the officer interrupted politely, "I promise you that we will send a police vehicle to transport you both to the Hermanus clinic. Meanwhile what your husband says is sensible. You need not worry. Your son is in no danger."

*

Across the road from the Hermanus Police Station was the two-storey medical clinic which served as the local hospital. Lexa stared at it in trepidation as the young policeman drove the yellow pickup van with police dogs barking loudly in the back. She noticed Soppie and Nelius outside the hospital *'non-whites'* entrance. Waving her hand out of the open van window in acknowledgement to them, they raised theirs back in return, before the vehicle turned from sight and she was dropped off at the *'whites only'* entrance around another side of the building. She hadn't seen Soppie or Nelius since leaving the barbecue at Joseph's party on Saturday night. Apprehensive, hands trembling, she pushed through the entrance swing-doors, frustrated at having to slow her step to that of a nurse who was waiting for her and who sedately led the way.

"How is Joseph, nurse?"

"He's recently come out of minor surgery. Everything should be fine. We've given him muscle relaxants and a local anaesthetic to set his wrist in plaster, with three fingers on that hand put into splints. Those fractured fingers are a mystery. The X-rays revealed bone compression as if his hand had been trapped under something heavy or badly squeezed. It's unlikely his fingers were fractured like that in a fall."

Lexa listened while they traversed a corridor.

"Apart from suffering concussion," the nurse continued, "surprisingly he only showed mild hypothermia after being exposed for quite some time in the cold ocean. His hand though was the worst injury. Apart from that he suffered a deep gash on his other arm which needed stitching plus a few bruises and minor abrasions. Otherwise, he was very lucky. You can have a few minutes with him now, just a few minutes, because he needs to sleep. There is the four o'clock visiting hour later."

They climbed the single flight of stairs. They approached a policeman sitting guard outside ward 7B. Lexa thought it odd to see a policeman on guard in the clinic. The officer spoke privately to the nurse before letting them through. A green nylon curtain was drawn around Joseph's bed. Lying inclined on his back with eyes closed, his torso was propped up so that the injured wrist set in plaster rested on a tray support. Seemingly asleep, Joseph appeared exhausted: his face gaunt and haunted, black rings under the eyes, pallid lips. There was a darkening yellow bruise on his shoulder and grazes up to the elbow of his good arm had been treated with splotches of red mercurochrome. As the nurse pulled the nylon curtain aside all the way round the bed, to brighten the enclosure, his eyes fluttered open. Lexa leaned over to him.

"Joe?" she whispered. Her tears dropped onto the white hospital sheet, which surprised her, as she knew she wasn't crying. She felt quite composed. He recognized her and half-smiled.

"Lexa.... they say I can check out of here soon." His voice sounded tired and groggy.

"Don't worry about that." She smiled, brushing a curl of his dark hair away from his forehead. "I won't disturb you for too long." She felt a flush of relief being with him. Her worst fears of his disappearance built-up and heightened in her mind over a terrifying sleepless night were now beginning to dissipate.

Suddenly an earnestness broke in Joseph's eyes.

"How is my father?" he said in a worried tone of voice trying to raise himself into a full sitting position, but was prevented by the nurse. "Do you know how my Dad is doing? Is he alive?"

Lexa looked at him with uncertainty.

"Fine Joe. They'll both be visiting you later."

He frowned as though confused.

"Did someone manage to get help for him? Did an ambulance reach my father in time?"

Unsure of what to make of these questions: perhaps his confusion was the influence of the muscle relaxant drugs or the after-effects of the local anaesthetic, she thought, or perhaps due to the concussion he had suffered or his exposure at sea. Lexa shrugged off these questions and smoothed down his sheet.

"Just rest yourself Joe," she reassured him, planting a light kiss on his forehead as the nurse began to draw the nylon curtain closed around the bed again, signaling her to leave. "We'll visit you later," Lexa said.

Joseph closed his eyes.

*

Downstairs at the clinic reception area, Lexa telephoned the Salem family to notify them of Joseph's injuries, the medical procedures he had undergone, and the four o'clock visiting hour. Having herself decided to wait the ninety minutes until visiting hour, she settled herself on a bench at the hospital entrance. It gave her time to consider the exchange of words which had occurred between Joseph's parents and the young policeman in the entrance hall at *Rus en Vrede*.

Why, she wondered, did Mr Salem request a police enquiry of Buck? Had she not known Buck so well, anyone would have thought they were apprehending a violent criminal the way that officer had truncheoned him around the police station this morning, literally assaulting him on the premises. All he seemed to be doing was passively resist arrest. Yet for a *coloured* man of Buck's stature under the State of Emergency regulations, resisting arrest was a major mistake, a hair trigger to unleash the wrath of the police.

But what crime had he committed? As far as she knew, Buck's life was not remarkable. He lived as a fisherman. But as an elder he sometimes took on the role of a spokesman for his people. Maybe that was it? Maybe his integrity and outspokenness, perhaps his educated confidence, baffled and frightened the police? Detained under section 29 of the Internal Security Act, so Sergeant Malan had explained to her in detail this morning, which meant Buck was arrested for subversive activities against the State. What a joke, she thought.

Lexa also remembered witnessing him being thrust into the police station still dripping wet. On the quiet - out of ear's range of the police - he had admitted to her having searched the kelp beds for Joseph, exhausting himself in the sea around *Voelklip*. But surely that in itself was not a political act. Nor was hiding Joseph's hanky which she herself had colluded in. It dumbfounded her sitting on the bench at the hospital entrance as to what politics Buck was arrested for. Particularly if Mr Salem was involved. Lexa decided to walk around to the *'non-whites'* entrance to talk with Soppie. She hoped that perhaps Soppie could provide more information about Buck.

"*Missie…*" said Soppie surprised to see her, standing up as she rounded the clinic. "We heard that Joseph was also brought to the clinic," he admitted.

"Thank goodness Joe is safe," she confirmed. "Apparently he slipped on the coastal path and fell off the cliff-face into the sea. They say he was swept away by the rip tide for kilometers along the shore. He had been missing for a day and almost two nights."

"He's lucky to be alive!"

She nodded. Then it surprised her how Soppie's words impacted on her emotions and Lexa found herself resisting tears from welling in her eyes. Soppie himself looked down miserably at the ground.

"Wasn't Nelius with you?" she asked.

"He's gone to call some friends."

"Is anything wrong?" Lexa asked.

Soppie sat down dejected on the stairs of the *'non-white'* entrance to the clinic. "It's Buck," he said.

"You know I saw him this morning being forced into the police station. Do you know why was he arrested?"

"It's far worse than that now."

"What do you mean, Soppie?" Lexa sat down beside him.

"Buck is no longer locked up in a cell in the police station. They've brought him over here."

"To the clinic. Why?"

"*Arme ou drommel*," Soppie uttered, covering his face with his hands. "Dear God have mercy. It is bad news, *Missie*. Nelius is warning the others and he has gone to call them over."

"But why?"

Supporting his forehead in his hands he shook his head from side to side as she watched him, waiting for him to speak. He rubbed his ears. "The police have reported that Buck is in an unstable critical condition." Soppie squinted across the road pointing vehemently at the police station. "Brought from over there!"

"You're joking! I don't believe it. But just this morning he looked fine..." Lexa closed her eyes feeling shocked at the implication.

"How much praying does it take, Missie?"

"Oh Soppie, I'm so sorry!"

For ten minutes they sat in a deadened silence thinking their own thoughts, meditating on their own anger, recalling their own memories of Buck. Until the faint approaching sounds of a crowd singing high in unison made them glance at each other. The sound emanated from the road which bypassed the *Coloured* Quarter. As the hidden sound of singing approached, the song became recognizable as the anthem *'Nkosi Sikelel iAfrika'*. But Soppie and Lexa maintained their silence, sitting side by side. There was little yet to be seen of the crowd. Except for the alarmed reaction of two young policemen across the road running frantically to their pickup, then speeding off in the direction of the distant song.

"I'll go in and ask again about Buck's condition," said Soppie. He entered the *'non-white'* section of the clinic as Lexa watched him through the glass doors approach a Muslim nurse at reception. The nurse listened to him, nodded, then walked off, returning after a few moments with a policeman. The officer spoke to Soppie with an accusing and pointed finger, Soppie shrugged, turned to the nurse who shook her head emphatically and pointed back to the policeman. Soppie threw his hands into the air with frustration and came sauntering back.

"What did they say?" Lexa asked.

"It's police business now." Soppie banged his fist angrily on the stair railing.

"Meaning?"

His face became very serious.

"Maybe Buck is dead," he said.

The crowd of singing men and women comprised an amalgamation of ordinary *coloured* and *black* town folk: house maids and chars, gardeners, labourers, shop assistants, school children and students, now visible in the main road with a yellow police van trailing close behind. The loudness of their song was deceptive. The crowd was a mere seventy to eighty strong, although to Lexa it seemed quite remarkable that Nelius could mobilize that many people on a workday at such short notice. It took

340

another five minutes before the group of singing protesters entered the parking-lot of the clinic, by which time three more police pickups, one carrying Sergeant Malan, were on the scene.

The two leading police vehicles screeched to a halt at the stairs where Soppie and Lexa were standing and in doing so, cordoned off the entrance. Eight policemen in camouflage outfits jumped from the vans all wielding riot gear and weapons including truncheons, quirts, teargas pistols. Two men who climbed onto a police van roof carried shotguns. Sergeant Malan strode up to the entrance landing with a megaphone in hand. At which point Lexa decided to move away from the stairs out of this possible line of confrontation. The chorus of singers arrived at the clinic *'non-white'* entrance, spreading themselves out as a group, and Soppie moved to greet them as their leader.

Songs of Africa wafted up from the parking lot. To Lexa the crowd who had gathered in front of her seemed a motley bunch. They were a random selection of Hermanus residents. They could hardly be called a mob against which the police were preparing to react with strong-arm tactics. Looking around, she could see their ages ranged from a baby being carried on its mother's hip to a wizened old fisherman propped up by a walking stick. A cohort of young workmen wearing blue overalls were *toyi-toying* by lifting their heels high in a dancing circle, one slow leg raised up after another as if they were tramping on hot coals. One workman with house paint still on his face after a half day on the job held up a paintbrush like an Olympic torch. While another man sporting a boater hat was trying with difficulty to strum his banjo to the tunes of the songs which were more anthems than carnival tunes. A group of students stood together arm-in-arm swaying in rhythm to a separate chant which they posed as a question:

"WHAT HAVE WE DONE? OH, WHAT HAVE WE DONE!"

Standing with these students was a wide-eyed child who grasped onto the hand of her older sister.

"Quiet please," Sergeant Malan interrupted, speaking into a megaphone. His amplified voice, however, was ineffectual as a roar went up from the group of *toyi-toying* workmen:

"Roar, young lions, roar - ROAR!"

"Viva ANC viva - VIVA!"

"Long live the ANC - LONG LIVE!"

"Roar, young lions, roar - ROAR!"

"Viva ANC viva - VIVA!"

"Long live the ANC - LONG LIVE!"

As the blockade of uniformed policemen waited nervously around their pickups, some with fingers ready on the triggers of riot weapons and firearms, Soppie toned his group into quiescence by signaling a hush with his outstretched arms. When he received acknowledgement, he turned to face Sergeant Malan.

"We have come peaceably to support our friend Buck in his hour of need," he said.

"This gathering is illegal," announced Sergeant Malan, now speaking too close into the megaphone with the result that his words distorted in a whine of feedback. He switched the device off, on, then repeated his announcement enforcing apartheid law. In reply, Soppie held his hands open before him in submission.

"We are peaceful," Soppie reiterated. "None of us here carry weapons. We just need to know what has happened to Buck Williams? His friends have come and they want to see him!" The crowd was stirred up by this statement; many people clapped in agreement with Soppie.

Standing above them at the top of the stairs, not quite the figure of an accomplished orator from ancient Greece, rather a hesitant police chief struggling to get sound out of a megaphone, Sergeant Malan softened his voice to a more reasonable tone.

"I'm afraid to say this meeting is illegal under the State of Emergency security regulations. All of you must disperse. We cannot allow a crowd to form around a public building. We will give you five minutes to disperse and go home."

Buck's crowd of friends in the parking lot moved forward towards Soppie who felt obliged to turn around to face them. A short debate ensued which after some discussion ended in a general consensus of opinion, some people nodding their approval, and the whole crowd sat down on the parking lot tarmac. Soppie remained standing up in front of them as their representative.

"We agree to wait here," Soppie shouted out. "To pray, to sing our songs, until at least I am allowed to visit Buck. Not one of us has seen him since he was detained in the police station this morning. We have heard rumours and we are worried about his health."

Sergeant Malan signaled to his closest uniformed man, issued an order to him, then watched the officer trot down to a van and speak into the receiver of a two-way radio. During this time the sitting crowd began to hum and sing softly in unison:

"WHAT HAVE WE DONE? OH, WHAT HAVE WE DONE!"

The policeman who had been issued with instructions returned to his commanding officer with a reply. Sergeant Malan nodded at the answer. Hesitantly he looked across at the crowd who were waiting.

One student shouted: "AMANDLA!"

"NGAWETU!" replied the group of workmen in blue overalls with sudden energy.

344

Two dozen fists from the crowd rose in black-power salutes.

Sergeant Malan ordered two of his men to arrest the student who had made the first subversive shout. Soppie intervened with the arresting policemen over this, and the crowd quietened. Waiting. On tenterhooks. It was clear that the police were getting nervous. Expectations were rising on both sides as Soppie looked up to Sergeant Malan who cleared his throat into the megaphone:

"Ahem... testing, testing... can you hear me... ja, that's better... This morning Mr Buck Williams injured himself seriously when, by accident, he fell down some stairs while in police custody. As you know he was detained under Section 29 for political questioning. We have police witnesses to confirm the accident. He died at two o'clock this afternoon."

The humming by the crowd seated on the tarmac stopped. Seconds passed before the impact of these last amplified words struck the collective heart of the group. At first, they seemed to have lost their voice, their unity of purpose. Then twenty students and workmen rose together to their feet as one united phalanx, and with them a booming, electrifying chant:

"AMANDLA - NGAWETU!"

A number of black power fist salutes went up.

"AMANDLA - NGAWETU!"

Black power salutes rose again.

"AMANDLA - NGAWETU!"

Before a teargas cannister landed dead-centre in the crowd exploding on the tarmac. Then another spiraled down a few feet away, obliterating the group in a blue pall of smoke for an instant. A shriek of pain went up as children started to yell and cry. Suddenly people were buckled over coughing, spluttering at the smoke, tears streaming from their eyes, their mucous membranes burning deep in the throat and in the nose. In their panic over this pain and disorientation the crowd scattered in all directions, dispersing away from the parking lot. And in a cynical mop up action a few policemen wearing gas masks gave up their guns for the chase with truncheons.

*

The uprising of Buck's friends had lasted less than fifteen minutes. Sickened by this confrontation: the intolerance of it, the armed might of the police, the terrible news about Buck - and feeling nauseous with the acrid smell of teargas still in her nostrils, Lexa sat shivering behind a clump of bushes. Just as the opposing parties had rallied forth, so she had backed off into a flowerbed and squatted down amongst the shrubbery to hide. And in the fray the police had forgotten about her.

Now, shaking with the shock of it, thirty minutes to go before Joseph's visiting hour, Lexa peered through the leaves of a bush as a last pick-up van exited from the parking lot and crossed to the police station over the road to offload people whom they had caught and arrested. She scanned the *'non-whites'* entrance: no one else was in sight except for the tail-end of a few of those people injured in the riot who were being ushered into the clinic under a police guard. The parking lot was otherwise quite deserted. It seemed strange to her how suddenly it was back to normal.

The only evidence remaining from the revolt were four scattered teargas cannisters and a few isolated pools of vomit. Two odd shoes were lying about, one without laces; there was an upturned hat and an unused disposable nappy and a jar of vaseline that someone may have used to smear onto their lips and eyelids to protect against the burning teargas. A gust of wind blew a white handkerchief across the tarmac.

Lexa waited with her legs drawn up to her chest, chin on her knees, eyes closed and streaming with tears caused by the teargas. Feeling scared to move out from the bushes, for a while she nodded off to sleep, not knowing for how long but waking and feeling less shaky. She waited until her breathing eased and she felt a desperate need to splash cool water into her face and eyes and drink some of it to ease her thirst and relieve the burning in her throat.

Negotiating her way around to the *'whites only'* entrance of the building, her eyes still burning and streaming with tears, Lexa wondered what this country was coming to. Is this what they mean by a nationwide State of Emergency? What they refer to as the black peril? Policemen who are terrified of a motley unarmed group of *coloured* and African men, women and children with the spirit of caring in the hearts for an old fisherman? Yes, and who are educated enough also to know about the power of free speech and democracy! With her vision blurred by the involuntary watering of her eyes, Lexa slowly trundled along into the entrance of the clinic and without too much effort she found the women's cloakroom.

*

At four o'clock Joseph's parents arrived at the Hermanus clinic. As Lexa greeted them at the entrance swing doors, Mrs Salem was in a state of anxious excitement. A door attendant wheeled Mr Salem along to an elevator as his wife was guided by the duty nurse at reception. They asked Lexa if she could wait behind, to come up to Joseph's ward in twenty minutes time. She could fully accept their need to be with their son in private.

When finally Lexa did climb the clinic stairs, she found the policeman still on guard seated outside ward 7B. Although Lexa asked, the policeman apparently knew less than her about the revolt she had witnessed in the parking lot. He did inform her, however, that until a police enquiry into Joseph's disappearance was completed and his political link to Buck was established, Joseph would be kept under surveillance.

Mrs Salem received Lexa warmly within the curtained enclosure around the bed, and Mr Salem smiled his smile as she sat down. Joseph was wide awake. Propped up against pillows. His appearance was much improved although he seemed reserved and very cool in the presence of his father - the only communication between them were hostile glances. Mrs Salem lightened the atmosphere by chatting about trivial things which she enjoyed. It put her in her element. She was obviously much relieved after her night's agony of waiting. Neither she nor Lexa knew of the events which had led to Joseph's unexpected dis-appearance. Mr Salem remained silent next to the bed.

Some minutes before the visiting hour ended Joseph and Lexa were given time alone together in the ward. Lexa had tried her utmost to involve herself during Mrs Salem's conversation, to act the part - to be light, carefree, happy - although she knew her efforts were unsuccessful. She simply could not function with the smell of teargas still powerful in her nostrils. Her hands were clammy, her body was shivering intermittently. But now alone with Joseph she was faced with the ambivalent feelings of being told not to stress him unduly yet needing desperately to talk to him about the riot she had just witnessed in the parking lot. And, of course, tell him about Buck.

They smiled at each other.

"You're looking better," Lexa said. To her the ward seemed to have grown much colder, the trembling of her body was getting worse. She felt a numbness in her reactions which she put down to shock.

"Have you read the latest news headlines?" Joseph asked. He passed her a newspaper. He pointed to an article. "Read that..." he said. "The government is now going to prevent the press from reporting to the world what is really happening here."

EMERGENCY LAWS TO BE TIGHTENED UP

State of Emergency regulations are to be redrafted to clear up or tighten up present definitions of what are deemed to be subversive actions.

In the process attempts are to be made to have more efficient, or tighter, control over news-papers through the Media Council, membership of which would become mandatory.

The Argus, Monday 8 December 1986

"Is that policeman still on guard outside my ward?" Joseph asked, lowering his tone of voice.

Lexa nodded.

"The police questioned me," he admitted. "They asked me about the contents of my journal. The police Special Branch have latched onto my *Hypothesis of*

Human Consciousness which I proposed in the journal from laboratory experiments. And actually..." he shook his head cynically. "It's absurd! They believe my scientific work is some kind of political document, a political manifesto or something." His eyes were sparked with incredulity. "They searched Buck's house after arresting him and discovered my pamphlet on FACT, an ecology group I proposed forming, and they warned me it could be incriminating evidence. But it is just an environmental group with a social awareness which we haven't even launched yet. My pamphlet on FACT and my journal have been confiscated by the police while they analyze them."

"Oh no, Joe. What about your years of hard work?"

"It's in here," he tapped his head. "I remember the journal results. If needs be I could transcribe them from memory. But if that experimental data is lost... it would take years to repeat all the experimental work." His eyes revealed a sudden anger as muscles bunched in his jaw. He tightened the fist of his good hand. "Buck must have been brought into the police station for questioning. Have you heard anything?"

Lexa swallowed but did not reply. She could not stop her hands from shaking, her body from shivering.

"What's wrong?" Joseph asked.

"It's nothing." She looked down at the floor.

"Come on, I can see there's something..."

Lexa started to sob quietly. She put her head down onto the bed, beyond his reach, so he was unable to touch or comfort her, held back by the straps around his injured arm. When at last she raised her head, her face was despondent, tear-soaked, shocked.

"He's dead Joe. Buck is dead."

Joseph looked coldly at her. In silence she stared at his disbelief. Then he lay his head back onto the pillows.

"You mean... our Buck?"

Lexa nodded.

He looked up at the ceiling repeating her words - Buck is dead - quietly to himself as if trying to squeeze some meaning out of them. She could see that he was still so drained from his ordeal in the sea, that the emotional impact of these words hardly seemed to grip his reality... they didn't find anchor anywhere in his feelings. Although a deep frown was incised on his face.

"How did it happen?" he asked.

"After they detained Buck early this morning, Sergeant Malan announced that he had tripped down a flight of stairs. They say the police can prove it was an accident."

"I suppose like accidentally slipping on a non-existent bar of soap, that old chestnut!" Joseph said in a biting tone of voice.

"How do you mean?"

"Never mind. Then what happened?"

"Well, a crowd of people from the *Coloured* Quarter had assembled outside in the parking lot of the clinic to hear news of Buck. When Sergeant Malan announced Buck's death a riot broke out."

"You're joking?"

"No. Look at me... Why do you think my eyes are red like this and streaming, my sinuses are stinging like mad and look how swollen they are. Just look at me. How I'm still shaking. The police used teargas..."

Joseph stared at her. He glanced back up at the ceiling. "It must have been Simon Thandiwe's motivation," Joseph murmured to himself while taking on the enormity of what had happened.

"They baton-charged the crowd," Lexa continued. "I saw old Dolfi amongst the injured who had been whipped and beaten by the police, and there were mothers involved with young children! The police attacked people indiscriminately."

"Are you hurt?"

"No, I'm alright, honestly. I was lucky."

Joseph's face had paled. "So Buck is really DEAD," he said with words full of self-reproach.

"It's not your fault Joe."

"Then who's fault is it? They discovered that pamphlet about FACT in Buck's house, with my name on it."

"It was your father..." Lexa admitted. "Your father was the one who reported him to the police. It wasn't your fault, Joe. As you say: FACT was just an environmental group!"

Joseph closed his eyes as his face whitened even more at the mention of his father. Now the full realization struck home. Buck had been sacrificed. Joseph held out his hand to Lexa.

"Come over here..." he said. As she walked around the bed, suddenly her emotions came gushing in a torrent and she buried her head in the white pillow shaking uncontrollably for a while as he held her with his good arm. In the quiet numbness which followed, the nurse appeared at the nylon curtain to inform them that the visiting hour was over. Lexa wiped her face. She heaved a sigh as she rested exhausted on his shoulder.

"We must just keep moving ahead," Joseph said, stroking her hair. "I know it's difficult, I know it seems that we're up against so much by taking this stance, by facing the truth like this, so let's try not to learn to hate."

She nodded.

"Dad was always threatened by Buck's outspokenness and by Buck's pride in who he was. Even at the party when Dad lied to us into thinking he was having a heart attack, Buck remained aloof. But like a fool I fell for his charade and nearly killed myself on that cliff-edge going to get help for him."

Lexa looked at him questioningly.

354

"Yes, my father actually faked a heart attack."

"You can't be serious, Joe!"

He nodded. "And the man still sits here next to my hospital bed without feeling even a tinge of guilt, without any inkling of an apology, not even thankful after I almost drowned trying to save his life which was never threatened in the first place."

"Is that what happened?" Lexa interlinked her fingers with Joseph's good hand.

"As you know Buck accompanied me at the party when it was time to confront Dad. But my poor father still can't see what the most important thing in my life is. Well, it's crystal clear to me now. And this weekend I've realized enough to give myself that freedom of choice." Joseph paused. "You know... so much freedom actually lies within us." These words seemed to come to him as a revelation. "Buck and I literally had to force my father to listen to my plans for the future. Can you imagine what Dad must have felt with Buck partly in charge! Can you imagine what he felt when I told him about the doctoral award!"

Joseph flexed his injured hand.

"I happen to be *white* and lucky. Buck wasn't."

They looked at each other.

"Find me a pen and paper quick," he said.

As he wrote Lexa watched him.

"One note is for Soppie: I'm requesting to have Buck's harmonica as I'm sure he would have wanted me to keep it. The other note is for my mother explaining what I have to do now." Joseph spoke in haste as the hospital intercom made a last call for visitors to leave the wards. "Lexa, will you go back to *Rus en Vrede* to pack an overnight bag for me. Give my mother this note: tell her not to worry and ask her to please feed my animals as usual. I'll collect everything else I need in a few days."

"Where are you going?"

"When they discharge me tomorrow, I'm leaving *Kwaaiwater* to begin the doctorate." Joseph's eyes seemed alive. "It will be straight back to my university life. Can you cope with that?"

She nodded. "What about your parents?"

"I've realized how impossible it is to move into the future by holding too firmly onto the past. My mother, I'm sure, will understand. Dad may take a lot longer to come around to appreciate my independence. It's not unlike those people you saw out in the parking lot today who are also now taking their lives into their own hands. Changing their expectations too fast now for anyone else to anchor the progress."

*

Walking along the pavement of the main road lined with aromatic shrubs and flowers, Lexa planned to phone for a taxi to return her to the Hermanus clinic once Joseph's overnight bag was packed. In her rush to vacate the ward after visiting hour, Joseph had little time to explain exactly what had occurred between him, Buck and Mr Salem, leading up to his accident on the cliff-edge. Or Mr. Salem's supposed heart attack at the birthday party. Or Joseph's interpretation of the cause of Buck's death. To her, there still seemed to be so many unanswered questions.

So, she thought back to this morning. To the clues she had tried to find in Joseph's bedroom. Searching for a particular object that was key. And she began to comprehend that it was not one or other specific object scattered in his laboratory room which held the answer. Instead, over the many years she had known Joseph, he had simply revealed the answer to her in the way he expressed his talents, in his lifestyle, in the passionate way in which he lived.

Lexa realized that you don't have to hide from your own voice. Your own talents need never become a curse or a bind. That by suppressing your own touch of genius, or the genius of others, you are suppressing the very heart of what makes you human. Ambling into the late afternoon light, she wondered how often she herself had made friends with that urge? How often had she actually lived to express as purely and truly as she could the voice inside her? Lexa imagined that if she tried to do this genuinely for herself, she may also begin for the first time to recognize properly that expression in others.

Joseph's handkerchief was still hidden in the pocket of her dress. To the touch it retained its dampness from the sea. She found comfort in holding onto it in her pocket as she walked slowly. She remembered Joseph and her once contemplating whether that prodigal baboon in his bedroom poster had made the right choice in facing the leopard head-on. Before now, she never believed that the baboon could survive such an onslaught, isolated on a desert salt pan in such circumstances...

But now she held out that hope.

The events she had witnessed in the past few days were proof of a triumph of will, achieved by making a choice and then taking a chance. Not that there was just one event involved in achieving such a triumph, she realized, but many events and many choices to be made: and so many people whom she knew that needed to take them up.

ABOUT THE AUTHOR

Brandon Broll is a distinguished poet listed in the *International Who's Who in Poetry and Poets' Encyclopaedia*. His latest poetry book is *Still-life of a Pandemic: In Three Books*. He is author of the bestselling science book *Microcosmos,* translated into multiple languages. Broll's fiction writing includes *London Bites: Eight Stories* and *Love in the Time of Brexit.* He lives in London and is married with two sons.